AF735038

Maize (*Zea mays* L.) in the Cold

Maize (Zea mays L.) in the Cold: Journey across Latitudes presents a compelling exploration of how one of the world's most vital crops has expanded from its tropical origins to endure and thrive in cold and temperate regions. As agriculture increasingly faces the challenges of climate variability and low-temperature extremes, this book offers a timely and comprehensive scientific account of maize's resilience to chilling stress.

Structured across seven in-depth chapters, the book takes readers through the **core biology of maize, its latitudinal journey from the tropics to cooler climates**, and a focused examination of **what constitutes low-temperature (LT) stress in maize.** It further explores the physiological, biochemical, and genetic **adjustments maize undergoes to cope with cold**, and highlights **emerging trends and scientific advancements** aimed at improving maize performance under such stress, including seed priming, molecular interventions, and breeding strategies.

Intended for researchers, maize breeders, plant biologists, agronomists, and postgraduate students, this volume also offers valuable insights for agricultural planners and policymakers working in cold-prone agro-climatic zones.

Dr. Salika Ramazan, a **low-temperature stress biologist** with a decade-long association with maize, brings together field expertise and academic depth to narrate the untold story of maize's adaptation across latitudes. Her work connects the scientific understanding of cold stress biology with practical innovations to strengthen maize production under challenging climates.

Maize (*Zea mays* L.) in the Cold

Journey across Latitudes

Salika Ramazan and Zahoor Ahmad Dar

CRC Press
Taylor & Francis Group
Boca Raton London New York

CRC Press is an imprint of the
Taylor & Francis Group, an **informa** business

Designed cover image: Salika Ramazan

First edition published 2027
by CRC Press
2385 NW Executive Center Drive, Suite 320, Boca Raton FL 33431

and by CRC Press
4 Park Square, Milton Park, Abingdon, Oxon, OX14 4RN

CRC Press is an imprint of Taylor & Francis Group, LLC

© 2027 Salika Ramazan and Zahoor Ahmad Dar

Reasonable efforts have been made to publish reliable data and information, but the author and publisher cannot assume responsibility for the validity of all materials or the consequences of their use. The authors and publishers have attempted to trace the copyright holders of all material reproduced in this publication and apologize to copyright holders if permission to publish in this form has not been obtained. If any copyright material has not been acknowledged please write and let us know so we may rectify in any future reprint.

Except as permitted under U.S. Copyright Law, no part of this book may be reprinted, reproduced, transmitted, or utilized in any form by any electronic, mechanical, or other means, now known or hereafter invented, including photocopying, microfilming, and recording, or in any information storage or retrieval system, without written permission from the publishers.

For permission to photocopy or use material electronically from this work, access www.copyright.com or contact the Copyright Clearance Center, Inc. (CCC), 222 Rosewood Drive, Danvers, MA 01923, 978-750-8400. For works that are not available on CCC please contact mpkbookspermissions@tandf.co.uk

For Product Safety Concerns and Information please contact our EU representative GPSR@taylorandfrancis.com. Taylor & Francis Verlag GmbH, Kaufingerstraße 24, 80331 München, Germany.

Trademark notice: Product or corporate names may be trademarks or registered trademarks and are used only for identification and explanation without intent to infringe.

ISBN: 9781041061519 (hbk)
ISBN: 9781041061533 (pbk)
ISBN: 9781003634027 (ebk)

DOI: 10.1201/9781003634027

Typeset in Times
by KnowledgeWorks Global Ltd.

Contents

Author Biographies

Dr. Salika Ramazan is a plant biologist with a Master's degree in Botany and a decade-long research focus on **low-temperature stress biology in maize**. She has qualified multiple national-level competitive exams, including CSIR-NET JRF, ASRB-NET, and JK-SET, reflecting her academic excellence and research aptitude.

Her work combines stress physiology, proteomics, and agricultural biotechnology to uncover how maize adapts to cold-prone environments, particularly in temperate and high-altitude regions. Dr. Ramazan has published her research in reputed international journals such as *BMC Plant Biology*, *Photosynthesis Research*, *Acta Physiologiae Plantarum*, and *Physiology and Molecular Biology of Plants*. Currently, she is pursuing postdoctoral research as a WISE Fellow, supported by the Department of Science and Technology (DST), Government of India. Her ongoing research continues to bridge the gap between molecular insights and practical approaches for enhancing maize resilience under cold stress.

This book reflects her deep commitment to the advancement of maize cold stress research and her contribution to climate-resilient crop science.

Dr. Zahoor Ahmed Dar is presently working as a Professor in SKUAST-K and is associated with AICRP—Maize Srinagar Centre as Incharge. He is involved in the development of 27 varieties in maize released at national and state levels. He and his team are credited with developing first QPM hybrid along with first drought-tolerant and first forage maize for the union territory of J&K. He has been associated with 30 externally funded projects from various funding agencies like DBT, DST, MoA, PPVFRA, UGC, NABARD, IFPRI, etc. He has developed 500 inbred lines which are in various stages of selfing, 24 populations, 12 pools, and 6 bi-parental populations in normal and QPM. He has deposited 90 maize accessions in the National Gene Bank at NBPGR and developed 20 Genetic Stocks for various traits. He has established heterotic groupings with contributing lines from Institute like VPKAS, Almora, IIMR, Ludhiana, and CIMMYT. He has published more than 300 papers in reputed journals.

Preface

Maize, one of the world's most essential food and feed crops, has a fascinating evolutionary and ecological journey—from its tropical ancestry in Central America to its wide cultivation in temperate and even high-altitude regions today. While the story of maize's adaptation to heat, drought, and nutrient challenges is well documented, its journey into the cold—both literally and scientifically—remains relatively underexplored.

This book, ***Maize (Zea mays* L.*) in the Cold: Journey across Latitudes***, is the culmination of my decade-long engagement with the chilling realities that maize faces in cold-prone environments, particularly in the Kashmir Himalayan and temperate regions. As a **low-temperature stress biologist**, I have witnessed firsthand the struggles of early seedling establishment under low temperatures, the physiological constraints that delay crop vigor, and the resilience strategies that maize exhibits when challenged by non-optimal thermal conditions.

The chapters in this book are organized to offer a coherent narrative—from the basic biology of maize and its latitudinal migration, to the mechanistic understanding of cold stress, and finally, to the innovations and interventions being developed to improve cold tolerance. Particular attention is given to how maize adjusts and adapts at cellular, physiological, and genetic levels, and to emerging solutions such as seed priming, genetic selection, and biotechnological advances.

This work is intended for plant scientists, maize breeders, agronomists, students, and agricultural practitioners who are either engaged in or curious about cold stress biology. I also hope it resonates with those working in climate-sensitive agro-ecosystems, where early season cold stress has become a critical barrier to productivity.

I am deeply grateful to my mentors, and the maize-growing communities of Kashmir and other temperate zones whose real-world challenges have shaped the urgency and direction of this work. I extend my heartfelt thanks to Taylor & Francis for recognizing the importance of this topic and supporting the publication of this volume. As maize continues its journey across latitudes and into a more uncertain climate future, may this book serve as a guide and a resource for those committed to resilient maize cultivation and agricultural sustainability.

Dr. Salika Ramazan
Srinagar, Jammu & Kashmir, 2025

Acknowledgments

First and foremost, I express my profound gratitude to Almighty God for His countless blessings, guidance, and strength that enabled me to complete this work. I would like to express my sincere gratitude to my mentor ***Prof. (Dr.) Zahoor A Dar,*** Dryland Agriculture Research Station (DARS), SKUAST-Kashmir for his invaluable guidance, constructive suggestions, and continuous encouragement throughout the preparation of this book. I would like to express my sincere gratitude to ***Prof. (Dr.) Manzoor A Shah***, Department of Botany, University of Kashmir for kindly providing laboratory space and access to research facilities that supported this work. I gratefully acknowledge the financial support provided by Department of Science and Technology, India (DST-IN) under the program DST PDF-WISE Scheme (DST/WISE-PDF/LS-131/2023), which significantly contributed to the successful completion of this work. My heartfelt gratitude to my PhD Supervisor, ***Dr. Riffat John***, Associate Professor, Department of Botany, University of Kashmir, for nurturing my intellect and guiding my scientific journey. Her mentorship and encouragement have shaped my academic foundation and inspired me to reach the stage where I am able to author a book independently today. Thank you so much, ma'am. I extend my heartfelt appreciation to my parents, ***Fahmeedah (Mother)*** and ***Mohammad Ramazan Naik (Father)*** for their unwavering support, inspiration, and blessings that have guided me at every stage of my academic and personal journey. I owe special gratitude to my husband, ***Dr. Shahzad A Pandith*** (Assistant Professor, Department of Plant Sciences, Central University of Kashmir), whose unwavering support, patience, and motivation have been central to this accomplishment. It was he who encouraged and inspired me to embark on the journey of authoring this book. His own scholarly achievements, including his authorship of a book (*Genus Rheum (Polygonanceae): A Global Perspective*) published by Taylor & Francis, have been a constant source of inspiration and confidence throughout this process. A special note of affection goes to my daughter, ***Haameem Binti Shahzad***, whose presence has been a constant source of joy and motivation. Finally, I acknowledge all those who have contributed, directly or indirectly, including my siblings ***Aneesa*** and ***Kashif Mohammad*** to the successful completion of this work.

Dr. Salika Ramazan

1 Introduction

Maize (*Zea mays* L.) stands among the most significant cereal crops worldwide, serving as a cornerstone of global food, feed, and industrial raw material. Owing to its exceptional genetic diversity, high productivity, and wide adaptability, maize has evolved from a regionally confined crop of the tropics to one cultivated across diverse agroecological zones, from equatorial lowlands to temperate high latitudes. This remarkable adaptability, however, has been accompanied by complex physiological, biochemical, and molecular challenges that continue to shape its productivity and resilience under changing environmental conditions. Originating in the tropical and subtropical regions of the Americas, maize was traditionally regarded as a warm-season, thermophilic C_4 plant requiring relatively high temperatures for optimal growth and development. Yet, through centuries of selection and modern breeding, it has extended its range to cooler temperate climates. This expansion has not been without physiological constraints, as low temperatures (LTs) can severely impair seed germination, seedling establishment, photosynthetic efficiency, and reproductive success. The process of acclimation and adaptation of maize to such environments therefore represents an extraordinary biological and evolutionary journey, reflecting the crop's ability to integrate structural, physiological, and molecular modifications in response to thermal stress. Understanding the biological underpinnings of maize adaptation requires a holistic appreciation of its taxonomy, reproductive biology, and biochemical composition. Beyond its agricultural relevance, maize has also drawn attention in pharmacological and industrial contexts, contributing to the global bioeconomy. However, with the increasing frequency of climatic fluctuations—especially temperature extremes—maize production is increasingly threatened by abiotic stresses such as drought, salinity, heat, and heavy metal toxicity. Among these, LT stress poses a particularly critical limitation in high-altitude and temperate regions, where it affects germination, early vigor, and ultimately yield.

This book, *Maize (Zea mays* L.*) in the Cold: Journey across Latitudes*, provides an integrated synthesis of the biology, ecology, and adaptive mechanisms of maize in the context of LT stress. Beginning with an overview of maize biology and its global cultivation patterns, the narrative progresses through its historical migration from tropical to temperate ecosystems, highlighting the physiological and molecular constraints encountered along this trajectory. Subsequent chapters explore the multifaceted responses of maize to LT; from morphological and anatomical adjustments to biochemical, physiological, and omics-level (genomic, transcriptomic, and proteomic) adaptations. The book also examines the contributions of modern molecular breeding and transgenic approaches toward enhancing cold tolerance, while situating these advancements within the broader framework of climate change and sustainable maize production.

DOI: 10.1201/9781003634027-1

By consolidating classical knowledge and recent research, this volume aims to offer a comprehensive understanding of maize's resilience mechanisms under LT stress. It serves as a valuable reference for researchers, agronomists, plant physiologists, and biotechnologists seeking to develop climate-resilient maize cultivars capable of sustaining productivity across diverse environments.

2 Biology of Maize

2.1 INTRODUCTION

Maize (*Zea mays*), commonly known as corn, is a member of the Poaceae family, which encompasses various grass species. Archaeological evidence and phylogenetic analyses indicate that maize domestication dates back at least 6,000 years (Matsuoka et al., 2002). This annual, erect grass typically grows to a height of 2–3 m, characterized by a distinct main culm with nodes and internodes. Maize holds a prominent position as one of the most vital cereal crops cultivated worldwide, owing to its versatility and extensive utility. Beyond serving as a staple food for humans and a primary feed source for livestock, maize plays a crucial role in numerous industrial sectors, contributing to products such as biofuels, starch, and sweeteners. Additionally, maize has gained significant recognition as a model organism in biological research, facilitating advancements in genetics, genomics, and plant physiology on a global scale. Maize (Z. *mays*) is among the most extensively cultivated crops worldwide, occupying over 100 million hectares (M ha) across various regions (Tubiello et al., 2009). As the global population is projected to rise from approximately 7.7 billion in 2020 to nearly 9.3 billion by 2050, maize demand in developing countries is expected to double within this period (Nur et al., 2020). Environmental conditions play a pivotal role in determining maize productivity, alongside its genetic attributes (Hudson et al., 2022). However, recent climatic shifts, including erratic rainfall patterns and temperature fluctuations, have posed significant challenges to maize cultivation. These changes have led to prolonged droughts and temperature extremes that deviate from optimal growth conditions, thereby impacting crop yields (Yadava et al., 2017; Odell et al., 2022).

In India, maize is cultivated across diverse agro-climatic zones, ranging from semi-arid to sub-humid and temperate environments. The ongoing climate crisis continues to reshape global agricultural landscapes, introducing further complexities to maize production. As of 2013–14, global maize production reached approximately 967 million metric tons (MMT), while India's contribution stood at 23 MMT (India Maize Summit, 2014). Maize, often referred to as the "Queen of Cereals," is renowned for its exceptional yield potential, making it one of the most significant cereal crops globally. The United States is the leading producer of maize, contributing approximately 35% of the world's total maize production. Often regarded as the "Mother Grain of Americans," maize plays a pivotal role in driving the U.S. economy. In India, maize cultivation is concentrated in key states such as Uttar Pradesh, Bihar, Rajasthan, Madhya Pradesh, Punjab, Haryana, Maharashtra, Andhra Pradesh, Himachal Pradesh, West Bengal, Karnataka, and Jammu and Kashmir, collectively accounting for over 95% of the nation's total maize output (Milind & Isha, 2013). In developing countries, the use of maize is variable. In Latin America and Africa, the

DOI: 10.1201/9781003634027-2

main use of maize is for food, while in Asia, it is used for food and animal feed. In fact in many countries, it is the basic staple food and an important ingredient in the diets of people. Globally, it has been estimated that approximately 21% of the total grain produced is consumed as food. It serves as a food crop for roughly 4.5 billion people in 94 developing nations, supplying about 30% of their daily caloric needs. In addition to being a significant source of oil, starch, biofuel, etc., 63% of maize grown worldwide is utilized as animal feed (Azadi et al., 2022). On the other side, Maize kernels are rich in nutrients, containing approximately 72% starch, 10% protein, 4% lipid, and various micronutrients such as vitamins and minerals (Nuss & Tanumihardjo, 2010). In recent years, increasing interest in nutraceutical development has highlighted maize's rich phytochemical profile and its potential health benefits also. As a result, maize-derived bioactive compounds have become a key focus in nutritional and medical research.

This chapter provides an in-depth discussion on the fundamental biology of maize, including its origin, evolutionary history, reproductive isolation, and genetic purity. It also delves into the biochemical composition and pharmaceutical potential of maize kernels. Furthermore, the chapter examines maize's ecological and environmental interactions, shedding light on current trends and future prospects, particularly in the context of both optimal and stress conditions amid changing climate. Additionally, it reviews global and Indian maize production patterns, along with its consumption dynamics, offering valuable insights into its significance in food security and sustainable agriculture.

2.2 TAXONOMY, ORIGIN, AND EVOLUTION

Maize (*Zea mays* L.) is a member of the tribe Maydeae within the grass family, Poaceae (Table 2.1)

The genus name *Zea* originates from the ancient Greek term "zela," which was used to describe a food grass. The genus *Zea* comprises four species, among which *Z. mays* is the most economically significant. The remaining species, collectively

TABLE 2.1
Scientific Classification of Maize

Kingdom	Plantae
Sub-kingdom	Tracheobionta
Super division	Spermatophyta
Division	Magnoliophyta
Class	Liliopsida
Sub-class	Commelinidae
Order	Poales
Family	Poaceae
Sub-Family	Panicoideae
Tribe	Andropogoneae
Genus	*Zea*
Species	*Mays*

known as teosintes, are wild grasses native to Mexico and Central America (Doebley, 1990). Maize has a chromosome number of 2n = 20.

The Maydeae tribe is divided into seven genera, classified into Old World and New World groups. The Old World group includes *Coix* (2n = 10/20), *Chionachne* (2n = 20), *Sclerachne* (2n = 20), *Trilobachne* (2n = 20), and *Polytoca* (2n = 20). The New World group consists of *Zea* and *Tripsacum*. Phylogenetic studies suggest that maize primarily evolved from the American genera *Zea* and *Tripsacum*, though *Coix* may have also contributed to its evolutionary history (Radu et al., 1997). The closest wild relatives of maize are the teosintes, all classified under the genus *Zea*. Outside this genus, *Tripsacum* represents the nearest relative. Among teosintes, Z. *mays* ssp. *mexicana* (Schrader), formerly known as *Euchlaena* mexicana or *Zea mexicana* (2n = 20), is considered the closest relative to modern maize. This annual teosinte, native to the central highlands of Mexico, is characterized by its large flowers and weedy nature, though it has limited use as forage or fodder (Doebley, 1990; Watson & Dallwitz, 1992). Other notable teosintes include *Zea diploperennis* and *Zea perennis*, which are diploid (2n = 20) and tetraploid (2n = 40) perennials, respectively, with restricted distributions in Jalisco, Mexico. Z. *diploperennis*, once on the verge of extinction in the late 1970s, has since become an important resource for maize ancestry research (Eubanks, 1995). Meanwhile, Z. *perennis* has been reported to have established a presence in South Carolina, USA (Hitchcock & Chase, 1951; Doebley, 1990). Another species, *Zea luxurians* (Durieu et Asch.), is an annual teosinte found in southeastern Guatemala (Doebley, 1990; Watson & Dallwitz, 1992).

The Mesoamerican region, encompassing present-day Mexico and Central America, is widely recognized as the center of origin for Z. *mays* (Watson & Dallwitz, 1992). Archaeological evidence indicates that maize domestication began at least 6,000 years ago, occurring independently in various locations, including the southwestern United States, Mexico, and Central America (Mangelsdorf, 1974). During the 16th century, Portuguese explorers facilitated the global spread of maize by introducing it to Southeast Asia from the Americas. Maize first arrived in Spain following Columbus's return from the New World, subsequently spreading to France, Italy, and Turkey. In the 17th century, the Portuguese introduced maize to India, from where it later reached China, the Philippines, and the East Indies. Today, maize is extensively cultivated in the United States, China, Brazil, Argentina, Mexico, South Africa, Romania, Yugoslavia, and India.

The domestication of maize has been the subject of multiple hypotheses (OECD, 2006). Several studies suggest that teosintes, particularly Z. *diploperennis* and Z. *mays* ssp. *mexicana*, along with species from the genus *Tripsacum*, played a crucial role in this process (Mangelsdorf, 1974). One early theory proposed that Z. *mays* ssp. *mexicana* originated from a natural hybridization event between *Tripsacum* and *Zea* (Mangelsdorf, 1974). Subsequent hybridizations between teosinte and wild maize are believed to have contributed to the development of modern maize varieties.

An alternative hypothesis suggests that either Z. *diploperennis* or *Tripsacum* may have interbred with an extinct wild maize species, giving rise to Z. *mays* (Purseglove, 1972; Radu et al., 1997). Further expanding on this idea, Eubanks (1993, 1997a) proposed that human selection of natural hybrids between *Tripsacum* and perennial teosinte played a pivotal role in the domestication of maize.

Maize is cultivated globally, and its genetic resources are preserved *ex situ* in various regions. However, only in the Mesoamerican region does *in situ* conservation of ancient maize persist—these original landraces serve as the genetic foundation for the improved varieties cultivated worldwide. The highest genetic diversity of maize is concentrated in the Mesoamerican region and the northern part of South America. The wide range of environmental conditions in these areas has contributed to the development of maize varieties that exhibit strong adaptability to challenging soil and climatic conditions, as well as resistance to biotic stresses. The diversity of maize is closely linked to cultural practices, agricultural production systems, and consumption patterns within local communities (Aguirre et al., 1998; Louette & Smale, 1998).

In many developing countries, there is a growing trend toward adopting improved maize varieties to meet market demands. However, this shift has led to genetic erosion. In Mexico, for instance, only 20% of the maize varieties cultivated 50 years ago are still in use today (World Watch Institute, 2000). The narrowing genetic base of modern maize underscores the critical need for conserving genetic diversity for future breeding programs. The International Maize and Wheat Improvement Center (CIMMYT) is at the forefront of maize germplasm conservation, maintaining the world's most extensive collection, with over 17,000 maize accessions.

India also possesses significant maize genetic diversity (Singh, 1977). Landraces with primitive traits, known as "Sikkim Primitives," are found in the North-Eastern Himalayan (NEH) region (Dhawan, 1964). Extensive germplasm collections from this region have been conducted by researchers at the Indian Agricultural Research Institute (IARI), New Delhi. Studies have shown that the two primitive maize strains from Sikkim (Sikkim Primitive 1 and Sikkim Primitive 2) differ from the primitive Mexican races (Mukherjee et al., 1989). Indian maize races have been classified into four groups: primitive, advanced (or derived), recent introductions, and hybrid races (Singh, 1977). The National Gene Bank in New Delhi safeguards approximately 6,000 indigenous maize accessions, and systematic evaluations are underway to identify agronomically valuable traits (Prasanna et al., 2009).

In addition to these traditional races, numerous local maize varieties exist in India. The genetic diversity observed today is a result of extensive hybridization between indigenous germplasm and foreign strains, particularly those from the United States (Mukherjee, 1989). Research has demonstrated that crosses between Indian maize varieties resulted in a 24–43% yield improvement, while hybrids between Indian and U.S. dent maize outperformed local varieties by 58% (Dhawan & Singh, 1961). Single-cross hybrids involving Indian and U.S. germplasm exhibited the highest yield potential, followed by U.S. × U.S. and Indian × Indian crosses, emphasizing the importance of genetic diversity for yield enhancement (Ahloowalia & Dhawan, 1963). Additionally, Dent × Flint hybrids, particularly those combining Indian and Caribbean, as well as Indian and U.S. germplasm, exhibited the highest heterosis, surpassing the better parent by 47–54% (Mukherjee & Dhawan, 1970).

2.3 BOTANICAL FEATURES AND REPRODUCTIVE BIOLOGY

Maize is a tall, determinate annual C4 plant with a height range spanning from less than 1 m to over 4 m. It develops large, narrow leaves arranged in an alternating, opposite pattern along a solid central stem (Figure 2.1)

The botanical characteristics of its various plant parts are as follows:

Maize plants typically develop three distinct types of roots:

1. **Seminal roots**—Emerging from the radicle, these roots persist for an extended period, supporting early plant growth.
2. **Adventitious (fibrous) roots**—Arising from the lower stem nodes below ground level, these serve as the primary functional roots, actively absorbing water and nutrients.
3. **Brace (prop) roots**—Formed at the lower two stem nodes, these roots provide additional structural support.

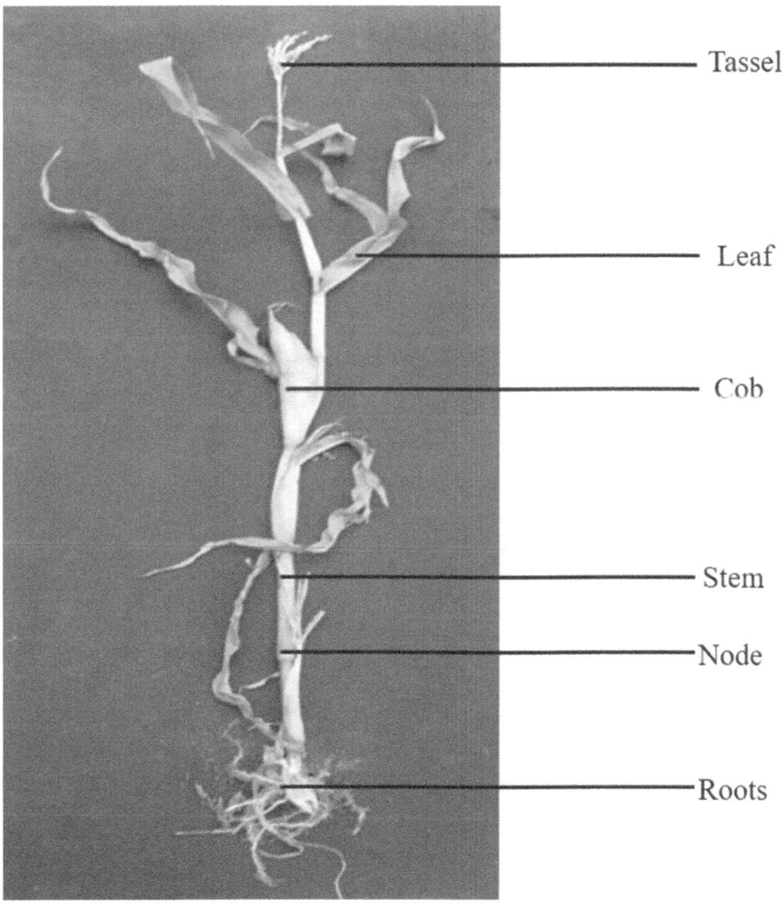

FIGURE 2.1 A maize plant.

Maize roots exhibit rapid growth, extending both outward and downward. Under optimal soil conditions, root expansion can reach depths and lateral spreads of up to 60 cm, facilitating efficient anchorage and nutrient uptake.

Stem: The maize stem typically grows to a thickness of 3–4 cm. It features short, robust internodes at the base, which progressively lengthen and thicken as they ascend the stem before tapering near the top. The internode bearing the ear (cob) is longitudinally grooved, ensuring proper positioning and support for ear development. The upper leaves play a crucial role in capturing sunlight and contribute significantly to the photosynthetic production that fuels grain formation.

Flower:

Maize is a monoecious plant, meaning it produces separate male and female inflorescences on the same plant.

- **Male Inflorescence (Tassel):** The stem apex terminates in a tassel, which consists of a loose panicle of male (staminate) flowers. Each spikelet within the tassel contains a pair of florets—one fertile and one sterile—enclosed within protective bracts.
- **Female Inflorescence (Ear or Cob):** The female flowers develop on condensed lateral branches known as shanks, which emerge from the leaf axils. The cob consists of a highly compacted rachis (central axis), with pairs of spikelets arranged along its surface (Figure 2.2).

The developing ear is tightly enclosed within multiple layers of husk leaves, remaining hidden until the emergence of pale yellow silk strands from the leaf whorl

FIGURE 2.2 Flowers in maize: male (staminate) and female (pistillate).

at the ear tip. These silks, which later turn green or purple, are elongated stigmas that serve as the receptive structures for pollination. Each spikelet contains two fertile florets, but only one typically develops into a maize kernel following successful fertilization by wind-dispersed pollen.

Due to the condensed internodes of the shank, the maize ear remains enclosed within its husks throughout development. Unlike wild grasses, maize has lost the ability to naturally disperse its seeds, relying entirely on human intervention for harvesting, seed extraction, and propagation.

Tassel:

The maize tassel is the male inflorescence, composed of multiple long, indeterminate branches that bear shorter, determinate spikelet pairs (Figure 2.3).

Each spikelet consists of two compact auxiliary branches, enclosed by two protective bracts, and contains a reduced male flower responsible for pollen production.

Stamen:

The stamen is the pollen-producing reproductive structure of maize, collectively known as the **androecium**. It consists of:

- **Filament (Stalk):** A slender structure that supports the anther.
- **Anther:** The terminal part of the stamen where pollen grains are produced and stored before dispersal.

FIGURE 2.3 Tassels (male flowers) in maize.

Pollen Development:

- **Microspore:** The smaller of the two types of spores produced in heterosporous plants, which matures within the pollen sac to form the male gametophyte.
- **Male Gametophyte:** Each microspore undergoes two divisions, forming a three-celled pollen grain (also called the pollen tube). This consists of:
  - **Two sperm cells**—One fertilizes the egg to form the zygote, while the other fuses with the central cell, initiating the development of the triploid endosperm.
  - **Vegetative (Tube) Cell**—Facilitates pollen tube formation, enabling sperm delivery to the ovule.

Reproduction and Propagation:

Maize reproduces exclusively through sexual reproduction, relying on wind-mediated pollen transfer for fertilization. Unlike some plant species, it does not propagate asexually in natural conditions. However, **cell and tissue culture techniques** can be employed for in vitro propagation, allowing the regeneration of maize tissues or plants from callus cultures.

2.3.1 Pollination and Fertilization

The formation of female flowers, or cobs, marks the first reproductive stage in maize and typically occurs 2–3 days after the tasseling stage. This stage begins when the silks emerge from the husk, signaling readiness for pollination. Unlike the tassel, which is a terminal inflorescence, maize ears develop as axillary flowers along the stem.

2.3.1.1 Pollination Process

Pollination occurs when airborne pollen grains land on the exposed, moist silks. Pollen shedding is not a continuous process—it generally begins 2–3 days before silk emergence and continues over 5–8 days, depending on environmental conditions. The silks are densely covered with fine, sticky hairs, which trap and anchor pollen grains for fertilization.

Pollen shedding is highly sensitive to humidity and temperature:

- Suspends during extreme conditions—Pollen release halts if the tassel becomes too wet or too dry.
- Resumes under favorable conditions—Shedding restarts once optimal temperature and humidity levels are restored.

Pollen remains viable for only 10–18 hours, with cool temperatures and high humidity prolonging its lifespan. Under ideal conditions, the anthesis-silking interval (ASI) lasts 1–2 days, but under stress, this period increases, potentially affecting seed set.

2.3.1.2 Fertilization Process

Fertilization begins once pollen adheres to the silk and germinates rapidly, forming a pollen tube that extends through the silk channel. This occurs within minutes of pollen contact. The pollen tube reaches the embryo sac in 12–28 hours, allowing fertilization to occur.

Pollen grains are lightweight and wind-dispersed, often traveling significant distances; however, most settle within 6–15 m (20–50 feet) of the source. Maize exhibits strong cross-pollination, with approximately 97% of kernels fertilized by pollen from other plants, rather than self-pollination.

2.3.1.3 Reproductive Isolation and Genetic Purity

Maize's natural tendency for cross-pollination necessitates reproductive isolation methods to maintain genetic purity, particularly for plant breeding and seed production. These methods include:

1. **Physical separation**—Maintaining sufficient spatial distance between different maize varieties to reduce cross-pollination.
2. **Physical barriers**—Using controlled pollination techniques or structural barriers to restrict gene flow.
3. **Temporal isolation**—Staggering planting times to prevent overlapping flowering periods between different maize varieties.

Hybridization between maize and related species—whether within the same genus (interspecific) or across genera (intergeneric)—requires the formation of a viable hybrid capable of producing fertile offspring. Even when sexual compatibility exists, the likelihood of successful hybridization depends on several key factors:

- **Proximity of species**—Closely related species have a higher probability of successful hybridization.
- **Environmental conditions**—Temperature, humidity, and wind patterns influence pollination efficiency.
- **Flowering synchronization**—Overlapping reproductive windows are essential for effective crossbreeding.

For gene introgression between maize and other species, multiple generations of recurrent backcrossing are often required to successfully transfer and stabilize desirable genetic traits.

2.3.2 Seed Dispersal

Unlike wild grasses, maize lacks a natural seed dispersal mechanism due to the structural characteristics of its ears. As a fully domesticated crop, maize has lost its ability to disseminate seeds independently and relies entirely on human intervention for propagation.

- Kernels remain tightly attached to the cob, preventing natural dispersal (Figure 2.4).
- If ears fall to the ground, high seed density leads to intense competition among emerging seedlings, greatly reducing survival chances and preventing natural establishment.

This dependency on human-assisted seed distribution is a defining feature of maize domestication.

FIGURE 2.4 Seed kernels on the cob in maize.

2.3.3 Corn Seed Anatomy

Grain: Botanically, the maize grain is classified as a caryopsis—a dry fruit in which the seed is fused to the fruit wall. The grain consists of two primary components:

1. **Germ**—Contains the plumule (embryonic shoot) and radicle (embryonic root), which develop into a new plant. The germ is also the source of maize vegetable oil, contributing 4% of the grain's total weight.
2. **Endosperm**—Provides nutrients for seedling growth until the plant establishes sufficient leaf area for autotrophic development. The endosperm makes up approximately two-thirds of the kernel's volume and accounts for 86% of its dry weight. It can be either yellow or white, depending on genetic variation (Figure 2.5).

FIGURE 2.5 Maize grains, caryopsis.

2.3.4 Anthesis-Silking Interval (ASI)

Grain number is a key determinant of maize yield, as in many other crops. Understanding the factors that enhance grain set—especially under drought conditions—is crucial for improving crop productivity. Research has shown that significant yield improvements can be achieved through selection under managed drought environments, even within existing elite germplasm. This suggests that genetic variation for drought tolerance exists but has remained largely untapped by previous breeding programs focused on yield potential and disease resistance.

In maize, drought stress from the mid to late vegetative stage does not alter the timing of tassel anthesis but delays ear silking. In drought-stressed genotypes, a prolonged ASI indicates poor drought tolerance, as drought-susceptible plants allocate fewer resources toward early ear development. This delay in silking leads to incomplete fertilization, causing grain abortion and reduced kernel numbers per ear.

To improve drought tolerance, breeders have employed recurrent selection for traits such as:

- Shorter ASI (indicating better synchronization of pollen shed and silking)
- Higher fertile ear number per plant
- Smaller tassels (reducing excessive resource allocation to male structures)
- Delayed senescence (enhancing resource allocation to grain filling)

These breeding strategies have led to greater partitioning of resources to early ear growth, ensuring successful grain set even under drought conditions. While poor timing of pollen shed can be an additional challenge (Du Plessis & Dijkhuis, 1967), natural plant-to-plant variation in pollen shedding duration often maintains sufficient pollen availability in maize fields.

Substantial genetic variation for ASI has been identified in tropical maize populations, particularly in Mexico. Recurrent selection in these populations—while maintaining yield potential and stay-green traits—has produced drought-tolerant germplasm that performs well across southern and eastern Africa (Banziger & Diallo 2004) and in drought-prone regions worldwide (Bruce et al., 2002). Recent studies confirm that drought tolerance in maize is largely additive, meaning that hybrid varieties can inherit drought resistance when both parental lines are drought-tolerant (Betrán, 2003).

Maize kernel morphology also varies across climates. Generally, drought-tolerant varieties produce kernels that are smooth and rounded, while ears tend to be long and slender with fewer kernel rows. In temperate zones, flint corn types tend to:

- Mature earlier
- Germinate better
- Exhibit stronger spring vigor
- Produce more tillers and fewer prop roots than dent strains

2.3.5 Mating Systems

Under natural conditions, maize reproduces exclusively through seed production. Pollination occurs when pollen from the tassel is transferred to the silks of the ear, enabling fertilization. Maize exhibits a strong tendency for cross-pollination, with approximately 95% of ovules fertilized by pollen from other plants, while 5% undergo self-pollination (Poehlman, 1959). Despite this, maize plants are fully self-compatible.

However, cytoplasmic male sterility (CMS)—a condition preventing self-fertilization—has been widely observed in maize. Rhoades (1931) and subsequent researchers have documented the occurrence of CMS, which is caused by mitochondrial genetic factors that disrupt pollen development. Based on differences in their restorer and maintainer reactionsand Beckett (1971) classified CMS in maize into three distinct types:

1. CMS-T (Texas)
2. CMS-S (USDA)
3. CMS-C (Charrua)

These CMS systems have been extensively utilized in hybrid seed production to facilitate controlled pollination and prevent self-pollination without the need for manual detasseling.

2.3.6 Methods of Reproductive Isolation

Maize, as a cross-pollinated crop, requires effective reproductive isolation methods to ensure the production of genetically pure seeds. To achieve this, plant breeders and seed producers employ various strategies to restrict gene flow. Common techniques include maintaining adequate separation distances between maize fields and implementing physical barriers to prevent unwanted cross-pollination. Additionally, several experimental practices are utilized to reinforce reproductive isolation and uphold seed purity in maize production.

i. **Maintaining isolation distance:** Cross-pollination in seed lots is managed by maintaining adequate separation between different maize lines. According to the Minimum Seed Certification Standards, an isolation distance of 200–600 m is recommended for maize seed production, depending on the variety and hybrid type (Tunwar & Singh, 1988). This practice effectively minimizes gene flow and helps maintain genetic purity.
ii. **Detasseling:** Mechanical detasseling is another effective technique in maize cultivation. By removing the tassel, which produces pollen and functions as the male flower, the potential source of genetic material that can be transferred via pollen is entirely eliminated. This method is widely used to control cross-pollination and maintain genetic purity in maize seed production.
iii. **Use of barrier crops:** The use of barrier crops has been recommended as an effective strategy to reduce isolation distances while still achieving the required separation for maize seed production. Barrier crops act as physical obstructions, limiting the movement of pollen by pollinators. African tall maize is commonly used for this purpose due to its greater height, dense growth, and ability to form an effective shield, thereby minimizing cross-pollination beyond the experimental plot. Consequently, incorporating appropriate isolation distances along with rows of African tall maize has become a standard practice in conducting confined field trials for genetically engineered maize.

2.3.7 Crossability between Zea Spp. and Hybridization

2.3.7.1 Intraspecific Crosses

Maize is predominantly a cross-pollinated crop species, with nearly 100% reliance on cross-pollination for reproduction. As a result, all maize varieties can freely interbreed, producing fertile hybrids (Purseglove, 1972).

2.3.7.2 Interspecific Crosses

Maize exhibits significant sexual compatibility with annual teosinte species, resulting in the formation of fertile hybrids (Wilkes, 1977). It has been reported that all teosinte species can cross with maize to produce fertile offspring, except for the tetraploid *Z. perennis*. However, maize-teosinte hybrids generally display low fitness, limiting their impact on gene introgression in subsequent generations (Galinat, 1988). The propensity for natural hybrid formation varies among teosinte species; for instance, *Z. luxurians* rarely hybridizes with maize, while *Z. mays ssp. mexicana* frequently produces hybrids (Wilkes 2004). Molecular evidence further confirms gene flow between maize and teosinte, indicating that introgression occurs bidirectionally, albeit at low levels (Doebley, 1990).

2.3.7.3 Intergeneric Hybridization

Although challenging, certain *Tripsacum* species, including *T. dactyloides*, *T. floridanum*, *T. lanceolatum*, and *T. pilosum*, have been successfully hand-crossed with maize to produce hybrids. However, these hybrids often exhibit a high degree of sterility and genetic instability. This infertility is common in wide crosses due to differences in chromosome number and the resulting lack of chromosome pairing (Eubanks, 1997b).

Maize-*Tripsacum* hybrids typically possess 28 chromosomes—10 from maize and 18 from *Tripsacum*—and are generally pollen-sterile with limited female fertility (Mangelsdorf & Reeves, 1939; Mangelsdorf, 1974). Research on Asiatic Maydeae species (*Coix*, *Sclerachne*, *Polytoca*, *Chionachne*, and *Trilobachne*) is limited, and no successful crosses with *Z. mays* have been reported. Isozyme analyses indicate that these Asiatic genera are genetically distinct from both maize and teosinte (Katiyar & Sachan, 1992). Although maize and *Coix* share the same chromosome number ($2n = 20$), chromosomal studies have revealed significant structural differences, suggesting limited potential for hybridization (Katiyar et al., 1992).

Nonetheless, species like *Trilobachne*, *Chionachne*, and *Coix* have been explored in germplasm screening programs to identify disease-resistant genes with potential applications in cultivated maize breeding (Sharma et al., 1995).

Maize readily crosses with hexaploid wheat (*Triticum aestivum*) and achieves high rates of fertilization and embryo formation. Plant breeders commonly utilize maize pollen to develop double haploids in hexaploid wheat. During the early stages of meiosis, however, maize chromosomes are eliminated from the developing genome, resulting in haploid embryos (Laurie & Bennett, 1986). There is minimal evidence to support the natural formation of fertile hybrids between maize and hexaploid wheat. Additionally, while there have been unconfirmed reports of hybridization between maize and sugarcane (*Saccharum* spp.), these claims lack substantial scientific validation.

2.3.8 Wild Relatives in India

In India, there are no sexually compatible wild or weedy relatives of *Zea* species present near maize-growing regions. The American genera, which include maize's close relatives, are not native to India. However, certain Oriental genera, particularly *Chionachne* species, are widely distributed across the country, especially in the arid regions of the Eastern and Western Ghats, Tamil Nadu, and other parts of India.

2.4 ECOLOGICAL AND ENVIRONMENTAL INTERACTIONS

2.4.1 Ecological Interactions

2.4.1.1 Outcrossing and Gene Flow

Gene flow from maize can occur through two primary mechanisms: pollen transfer and seed dispersal. However, seed dispersal in maize is relatively easy to control, as domestication has largely eliminated the natural dispersal mechanisms that ancestral maize once possessed (Purseglove, 1972). Maize kernels are tightly attached to the cob, and if the ear falls to the ground, competition among emerging seedlings restricts their growth to maturity (Gould, 1968).

Pollen movement remains the primary means of gene escape from maize plants. Since maize is predominantly cross-pollinated, factors such as wind speed and direction significantly influence pollen distribution. Maize pollen, measuring approximately 0.1 mm in diameter, is the largest among grass family members and is dispersed by wind from a relatively low elevation. Due to its larger size, maize pollen settles roughly ten times faster than pollen from other wind-pollinated species (Di-Giovanni et al., 1995). Studies by Raynor et al. (1972) demonstrated that maize pollen travels shorter distances by wind compared to smaller pollen grains, exhibits limited horizontal and vertical dispersal, and tends to settle quickly, with much of it remaining within the vicinity of the source plant.

Insects such as bees have been observed collecting pollen from maize tassels; however, they play a minimal role in cross-pollination since they lack an incentive to visit the female flowers (Raynor et al., 1972). In commercial maize cultivation, slight differences in flowering dates between varieties may still allow cross-pollination if fields are adjacent.

The limited viability of maize pollen further reduces the risk of cross-pollination, as pollen must encounter a receptive host within approximately 30 minutes of being released to remain biologically active (Luna et al., 2001). Additionally, cross-pollination likelihood is influenced by the concentration of maize pollen in the environment; maize pollen produced in higher volumes can outcompete foreign pollen sources (Raynor et al., 1972).

Gene flow from maize (*Z. mays*) to other species within the same genus (interspecific) or across genera (intergeneric) depends on several conditions. First, a viable intermediate hybrid must form and produce fertile progeny capable of surviving into subsequent generations. Even with sexual compatibility, hybridization success is influenced by factors such as the proximity of maize to related species, environmental conditions, and overlapping flowering periods. Successful introgression of maize genes into other species may require multiple generations of recurrent backcrossing.

2.4.2 Potential for Gene Transfer from Maize

2.4.2.1 Gene Transfer between Different Maize Species

Available evidence indicates that *Z. mays* can successfully cross with teosintes, producing fertile hybrids. However, these hybrids generally exhibit low fitness, resulting in minimal impact on gene introgression in subsequent generations. While the similarity in chromosome number suggests a potential for hybridization between *Z. mays* and wild species from the Asiatic genera (*Coix*, *Sclerachne*, *Polytoca*, *Chionachne*, and *Trilobachne*), research on these genera remains limited. To date, no confirmed reports have documented successful crosses between *Z. mays* and these Asiatic species.

2.4.2.2 Gene Transfer from Maize to Other Plants

The introgression of genetic information from one plant species to another holds significance only when the two species are sexually compatible and their hybrid offspring are viable. In the case of maize, this process is generally not applicable, as successful gene flow through hybridization is highly restricted due to limited sexual compatibility and the low viability of resulting hybrids.

2.4.2.3 Gene Transfer from Maize to Other Organisms

Horizontal gene transfer from plants to animals, including humans, or to microorganisms is considered extremely unlikely. There is no known mechanism by which maize genes—or genes from any other plant—could be transferred to humans or animals. Despite the extensive consumption of plant DNA over evolutionary history, no evidence has been identified to suggest that such gene transfer has occurred. Consequently, the likelihood of maize genes integrating into human or animal genomes is effectively zero.

Similarly, gene transfer from maize, or any other plant, to microorganisms is highly improbable. Horizontal gene transfer from plants to bacteria has not been demonstrated under natural conditions (Nielsen et al., 1997, 1998; Syvanen, 1999). Furthermore, deliberate experimental attempts to induce such transfers have been unsuccessful (Schlüter et al., 1995; Coghlan, 2000).

2.4.3 Free-Living Populations of Maize

The term "free-living" refers to plant populations capable of surviving independently over the long term, competing successfully with native flora without direct human intervention. This ecological category includes plants that establish themselves in open, disturbed environments, whether in human-managed spaces (weedy populations) or naturally disturbed areas such as riverbanks and sandbars (wild populations). In India, no such free-living populations of maize have been reported.

2.4.4 Volunteers and Weediness

As previously mentioned, maize has acquired agronomically significant traits through domestication, making it highly dependent on human intervention for seed dissemination. Due to this extensive domestication, maize seeds remain tightly attached to the cob and cannot disperse naturally without human involvement.

Although maize seeds from the previous crop year may occasionally germinate the following season, maize cannot establish itself as a persistent weed. While volunteer maize plants are common in some agronomic systems, they are easily managed since maize lacks the ability to sustain reproduction outside cultivated environments. Maize is considered non-invasive in natural habitats (Gould, 1968).

2.4.4.1 Environmental Interactions

Environmental factors play a crucial role in determining maize crop yield. Maize productivity and associated traits are influenced by both genetic factors and environmental conditions (Hudson et al., 2022). While plants naturally progress through defined growth stages to complete their life cycle, recent climate changes—including erratic rainfall patterns and temperature fluctuations—have posed significant challenges to crop production. Extended droughts and temperature shifts beyond optimal ranges have further exacerbated these issues (Yadava et al., 2017; Odell et al., 2022).

In India, maize is cultivated across diverse environments, ranging from semi-arid to sub-humid and moderate climates. Climate change remains an ongoing global concern, causing substantial environmental shifts that threaten agricultural stability. Although crop productivity has improved in recent decades, rising vulnerability to abiotic stresses presents a major obstacle to maintaining high yields under changing climate conditions (Mao et al., 2015; Alotaibi, 2023). Developing maize varieties with enhanced abiotic stress tolerance is increasingly recognized as essential for future food security (Duvick, 2005). In response to environmental stressors, plant cells activate intricate signaling pathways involving hormones, transcription factors, and signal transducers. These coordinated responses regulate stress-responsive genes, prompting the production of protective proteins and enzymes that aid in stress adaptation (Zandalinas et al., 2018).

Maize production faces significant risks from drought, extreme temperatures, and inconsistent rainfall patterns (Lobell et al., 2014). Consequently, ongoing research is dedicated to improving maize traits that enhance resilience to such abiotic stresses. However, identifying the genetic determinants of stress tolerance remains a complex challenge. Abiotic stress tolerance is governed by intricate quantitative traits, often linked to developmental processes, and controlled by multiple quantitative trait loci (QTL), each contributing marginally to overall trait expression (Miao et al., 2017). Using plant-supporting microorganisms such as arbuscular mycorrhizal (AM) fungi, various other beneficial fungi, and plant growth-promoting bacteria (PGPB) presents a promising alternative strategy (Dodd & Ruiz-Lozano, 2012; Barea, 2015). These microorganisms can enhance crop yields without the need for additional mineral nitrogen. Research on maize has highlighted their substantial benefits, including improved growth and better crop quality (Montañez et al., 2012; Berta et al., 2014; Dhawi et al., 2015; Marks et al., 2015; Dartora et al., 2016). Employing PGPR is a promising strategy to mitigate the environmental effects of chemical fertilizers, pesticides, and herbicides, as well as to address abiotic stress. According to Kloepper (1978), PGPR are soil bacteria that colonize the plant rhizosphere and support growth through various mechanisms. With the increasing demand for agricultural output to support a growing global population, excessive reliance on chemical inputs has often led to soil degradation (Bhusal et al., 2021). Plant growth-promoting microbes have been shown to boost plant nutrition and reduce the need for pesticides (Kimotho et al., 2019). With modern agriculture grappling with environmental and social issues from industrialization and the pressure to feed a growing population, PGPR presents a viable solution for sustaining high yields while reducing environmental impact (Pérez-Montaño et al., 2013).

Researchers are increasingly exploring microbial strategies to enhance maize's resilience against both biotic and abiotic stresses. Maize crops are vulnerable to damage from insect pests and microorganisms during both pre-harvest and post-harvest stages, resulting in significant reductions in quality and yield (Burlakoti et al., 2024). Pathogenic bacteria and insects collectively contribute to an estimated 9.4% annual global loss in maize production. Among these factors, insect pests are the primary cause of maize degradation and yield decline, while fungi serve as a secondary contributor (Khosravi et al., 2007; Sitara & Akhter, 2007). Pests—particularly insects and soil-borne pests—pose considerable challenges to maize cultivation. Global losses due to plant diseases, including both direct damage and indirect impacts on crop productivity, are estimated to reach approximately $40 billion annually (Roberts, 2006). This highlights the urgent need for improved pest management strategies, with microbial approaches emerging as a promising solution for safeguarding maize yields.

2.4.5 Environmental Factors

Abiotic and biotic factors are key environmental determinants of crop productivity, and their effects are increasingly intensified by rising global temperatures, further exacerbating climate change. Abiotic stressors such as cold, drought, heat, salinity, and flooding adversely affect plant growth and productivity by inducing a range of physiological, morphological, molecular, and biochemical changes. These stressors also alter soil conditions, impacting its composition, pH, and physicochemical as well as biological properties.

Biotic factors include both beneficial organisms, such as decomposers, natural enemies, and pollinators, and harmful agents like weeds, pathogens, arthropods, vertebrates, and human-induced disturbances (Burlakoti et al., 2024). Both abiotic and biotic stressors pose significant challenges to sustaining crop yields, highlighting the need for adaptive strategies to mitigate their impacts (Figure 2.6).

Maize production is significantly affected by biotic stress, primarily caused by diseases and insect pests (Lodha et al., 2013). Common maize diseases include aflatoxin contamination, sugarcane mosaic disease, ear rot, northern leaf blight, and rough dwarf disease. Among the major insect pests are corn borers from Europe, the Mediterranean, and tropical regions, as well as the maize weevil, which causes substantial damage to stored grain.

In the United States, maize anthracnose, caused by the fungus *Colletotrichum graminicola*, results in losses estimated at up to $1 billion annually (Frey et al., 2011; Balmer et al., 2013). Globally, stalk-boring lepidopteran larvae inflict considerable economic harm, with the European corn borer (ECB) posing a particularly severe threat. ECB larvae feed on kernels and burrow into maize stalks, often causing plant lodging during harvest.

In tropical and subtropical regions, maize weevils (*Sitophilus zeamais* Motsch) are especially problematic, frequently causing grain losses exceeding 30% during storage (Tigar et al., 1994). This extensive pest prevalence underscores the need for effective management strategies to safeguard maize productivity.

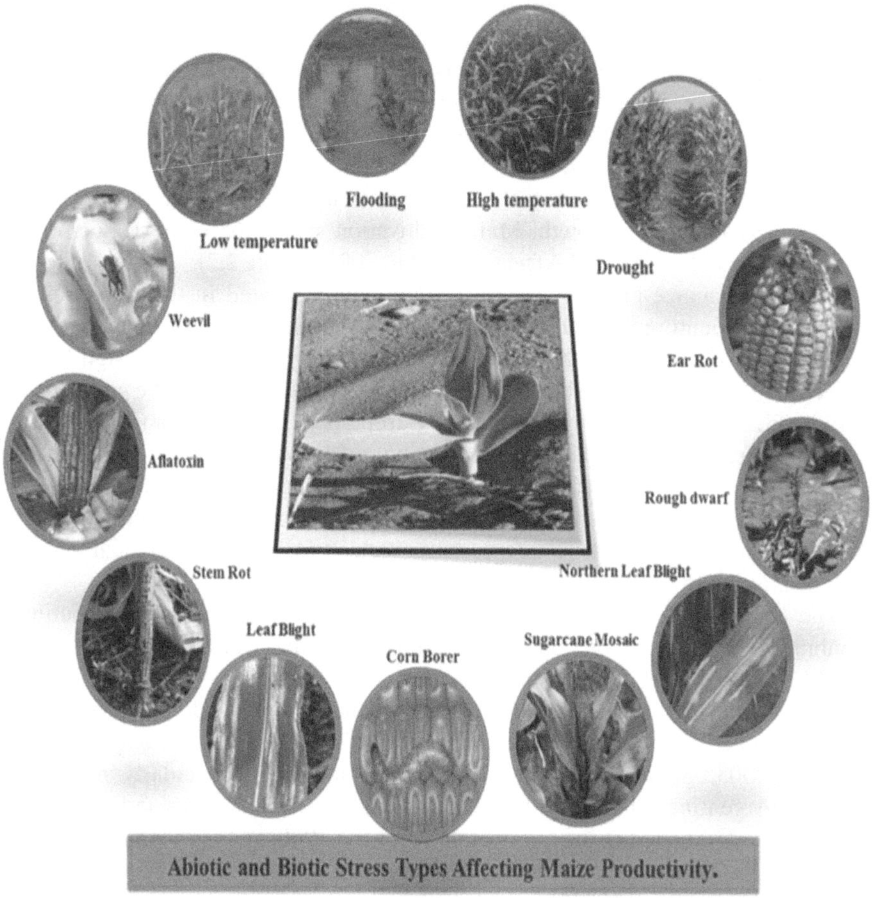

FIGURE 2.6 Abiotic and biotic factors affecting maize.

2.5 MAIZE CULTIVATION: A GLOBAL SCENARIO

2.5.1 Trends in Global Maize Production

Maize is cultivated annually on approximately 197 M ha worldwide for dry grain production, making it the second most widely grown crop globally after wheat. In comparison, wheat occupies around 216 M ha, while rice covers approximately 165 M ha. Despite being second in cultivation area, maize leads in global dry grain production, reaching 1,137 million tons (M t) annually (TE2019), which is about 50% higher than the production of both rice and wheat (757 M t each). This production advantage is largely attributed to maize's higher average grain yield of 5.8 tons/ha, driven by extensive hybrid cultivation and enhanced input use. Over the past 25 years, maize production has more than doubled (+118% since TE1995), supported by both substantial yield improvements (+50%) and expanded cultivation. Since 1961, the global maize cultivation area has increased significantly, rising from

106 M ha (TE1963) to the current 197 M ha, with a notable acceleration in expansion since the early 2000s.

Maize is on track to surpass wheat as the most widely cultivated crop by 2030 if current trends continue, as wheat cultivation areas have remained relatively stagnant (Erenstein et al., 2021). Since 1961, global maize yields have nearly tripled, rising from 2 tons/ha (TE1963) to 5.8 tons/ha (TE2019), marking a 190% increase. Consequently, maize production has surged fivefold over the same period, reflecting a remarkable 441% growth. Maize cultivation spans 165 countries across the Americas, Asia, Europe, and Africa (FAOStat, 2021) (Table 2.2).

The global maize-growing area is predominantly located in the Americas and Asia, each accounting for over one-third of the total, followed by Africa with roughly one-fifth and Europe with about one-tenth. Regional yield variations significantly influence maize production shares. The Americas contribute half of the global maize output (TE2019), followed by Asia (32%), Europe (11%), and Africa (7.4%). Within these regions, there are notable differences in production patterns. In the Americas, maize cultivation is split between North America (primarily the United States) and Central/South America. In Asia, two-thirds of the maize area is concentrated in East Asia (primarily China), where temperate maize dominates, while tropical maize is prevalent in South and Southeast Asia.

The United States and China are the leading maize producers, contributing a combined 54.5% of global output (TE2019), with the United States producing 361 M t and China 259 M t annually. Eight major producers—the United States, China, Brazil, Argentina, Ukraine, Indonesia, India, and Mexico—each produce over 25 M t per year, collectively accounting for 77.4% of global maize production.

Since 2003, global maize area expansion has been particularly notable in Ukraine, Argentina, China, and Indonesia. While the United States and China have consistently dominated maize production since the 1960s, China's contribution has

TABLE 2.2
Global Corn Production 2024/2025

Country	% of Global Production	Total Production (metric tons)
United States	31%	377.63 Million
China	24%	294.92 Million
Brazil	10%	126 Million
European Union	5%	58 Million
Argentina	4%	50 Million
India	3%	38 Million
Ukraine	2%	26.5 Million
Mexico	2%	23.7 Million
South Africa	1%	17 Million
Canada	1%	15.35 Million

Source: USDA (Foreign Agriculture Service, U.S. Department of Agriculture), accessed on March 4 2025; https://www.fas.usda.gov/data/production/commodity/0440000

increased significantly—rising 14-fold—whereas the United States' output grew nearly fourfold. The surge in U.S. maize production from the early 2000s to 2010 was largely driven by ethanol biofuel demand, with yield improvements playing a crucial role rather than significant expansion in maize cultivation area (Wallington et al., 2012). Historically, the top maize producers in the 1960s included the United States, China, USSR, Brazil, Mexico, South Africa, Romania, and Yugoslavia, collectively accounting for 76.4% of global maize production (TE1963). Maize's adaptability to diverse agro-ecological environments—ranging from wet to dry conditions and from lowlands to highlands—has led to the classification of distinct rainfed maize mega-environments. These environments are defined based on climatic factors such as maximum temperatures and rainfall, facilitating targeted breeding efforts to improve germplasm for specific production zones (Bellon et al., 2005).

The concept of maize mega-environments plays a crucial role in defining the types of maize cultivated and the traits required for improved adaptability. These mega-environments influence maize selection for specific conditions, such as temperate or tropical climates, short- or long-season growth cycles, and tolerance to drought, heat, and biotic stress (Prasanna et al., 2021). In regions like Kenya, maize cultivation is classified based on altitude: tropical lowlands (0–1,000 meters above sea level (masl)), mid-altitude areas (1,000–1,800 masl), and long-season highland zones (>1,800 masl). However, while temperature and rainfall remain key factors in defining these zones, this agro-ecological framework overlooks other important variables such as soil type, topography, and socio-economic conditions like farmer practices and market access (Bellon et al., 2005). Climate change is expected to significantly alter these maize mega-environments. Increased warming is likely to expand maize cultivation into northern and southern latitudes and higher altitudes. However, this shift is accompanied by a greater frequency of abiotic stresses like heat, drought, and waterlogging, as well as heightened risks from biotic threats such as diseases and insect pests (Jones & Thornton, 2003; Cairns et al., 2013; Tesfaye et al., 2015; Prasanna et al., 2021).

In Africa, recent years have seen devastating outbreaks of maize lethal necrosis (MLN), a transboundary disease, and the fall armyworm (*Spodoptera frugiperda*), which invaded over 40 African countries since 2016 and subsequently spread to Asia in 2018 (Prasanna et al., 2018; Boddupalli et al., 2020; De Groote et al., 2020; Kassie et al., 2020; Li et al., 2020). Climate change is anticipated to intensify these biotic threats, fostering the emergence of new pests and diseases (Deutsch et al., 2018; Burdon & Zhan, 2020). On a global scale, more than two-thirds of maize cultivation relies on conventionally improved, non-GM varieties. However, maize has become the second most widely grown genetically modified (GM) crop worldwide, occupying 32% of the total GM crop area (after soybean at 48%). As of 2019, GM maize covers approximately 61 M ha, representing 31% of total maize cultivation (ISAAA, 2019).

Insect-resistant Bt maize has provided significant benefits, especially in the United States, Brazil, South Africa, Canada, and Argentina—with its impact comparable to Bt cotton (Brookes & Barfoot, 2020). Herbicide-tolerant maize, mainly linked to glyphosate resistance, has delivered substantial cost savings and yield improvements, particularly in the Global South. While GM drought-tolerant maize adoption remains limited, it is primarily cultivated in the United States (Brookes & Barfoot, 2020). In Southeast Asia, GM maize adoption is growing steadily. The Philippines became the first Asian country to commercialize GM maize, introducing Bt yellow maize in 2003, which was later enhanced with herbicide-tolerant traits (Afidchao et al., 2014).

2.5.2 Trends in Global Maize Consumption

Maize is a highly versatile crop with multiple uses across the globe. While maize grain is predominantly used as animal feed (56% of total production), a significant portion (20%) is allocated for non-food purposes, and only 13% is directly consumed as food. However, this classification underestimates maize's overall contribution to human nutrition. A substantial part of maize's impact on food systems comes through indirect consumption pathways. For instance, maize used as animal feed contributes to the production of animal-sourced foods such as meat, milk, and eggs. On average, producing 1 kg of boneless meat requires approximately 3 kg of human-edible feed (mainly maize grain and soy)—with ruminant systems requiring 2.8 kg and monogastric systems about 3.2 kg (Mottet et al., 2017). Additionally, maize by-products like stover and forage/silage maize provide vital feed resources for livestock. Maize grain ranks as the third most consumed cereal in the world after rice and wheat in terms of direct human consumption. Global maize consumption for food averages 18.5 kg per capita annually, accounting for 11% of the total average cereal intake of 175 kg per capita (2014–2018 data, FAOStat, 2021). Maize's role as a staple food is more prominent in low- and lower-middle-income countries, where 43% of maize production is directly consumed. In Africa, this figure rises to 54%, with eastern and southern Africa showing particularly high dependency at 66%. Consumption levels exceed 50 kg per capita annually in 22 countries, notably in parts of eastern and southern Africa (including Lesotho, Malawi, Zambia, and South Africa) and Latin America. In these regions, maize serves as a primary staple, with some countries reporting consumption levels exceeding 100 kg per capita per year. Mexico, with its rich maize-based culinary traditions, is also among the leading consumers.

Over the past decades, global per capita maize food consumption has steadily increased from 11 kg (TE1963) to 18.8 kg (TE2018). However, aggregate maize use has grown even faster, driven by population growth in Africa and Asia, along with increased maize utilization for feed, industrial, and energy purposes (e.g., in the United States; Kumar & Singh, 2019). The global population is projected to rise from 7.7 billion to 9.7 billion by 2050 (ranging from 8.9 to 10.7 billion depending on fertility rates; UN-DESA, 2019). Assuming constant annual per capita maize food consumption, this growth could demand an additional 37 M t of maize annually for food by 2050, with estimates varying between 22 and 56 M t based on fertility assumptions (Table 2.3).

Maize's diverse applications stem from its multiple types, differing in color (yellow, white, blue) and attributes (e.g., dent/flint, sweet corn, baby corn, popcorn, waxy maize, high-amylose maize, high-oil maize, and quality protein maize (QPM); Serna-Saldivar & Perez Carrillo, 2019). These variations reflect differences

TABLE 2.3
Global Maize Consumption

Country	Average Use (% of domestic supply)					
	Food	Feed	Seed	Post-Harvest Losses	Processing	Other Uses (Non-food)
Africa	54.3	30.3	1.1	8.3	1.6	3.5
Asia	11.6	64.2	1.0	4.9	2.7	15.6
Americas	7.5	50.5	0.5	6.9	5.9	28.7
Europe	5.6	75.7	0.9	2.6	9.0	6.4
Oceania	18.5	66.8	0.2	1.5	3.1	10.0
Low- and lower-middle-income countries	43.1	41.0	1.7	7.9	1.6	4.9
Upper-middle- and high-income countries	7.6	59.0	0.6	5.5	5.2	22.1
World	12.8	56.3	0.7	5.8	4.7	19.6

Source: FAOstat, 2021.

in endosperm texture, pericarp structure, starch type, and kernel characteristics (García-Lara et al., 2019).

- White maize is primarily used for food.
- Yellow maize is largely used for animal feed.
- Blue maize is typically linked to niche food markets and industrial applications (Keleman & Hellin, 2009; Blare et al., 2020).

Maize's role in human diets varies widely—from fresh green maize (kernels on the cob) consumed seasonally to processed ingredients and staple foods. Approximately 5% of global dietary calories and proteins are derived from maize grain. On average, maize contributes 156 kcal to the global daily per capita energy intake, out of a total of 2,919 kcal. However, maize's contribution to energy intake is significantly higher in Africa and Latin America, exceeding 400 kcal per capita in certain regions. In eastern and southern Africa, maize contributes nearly 25% of total dietary energy. Protein intake from maize as food averages 3.8 g per capita daily, accounting for 12% of cereal-derived protein intake globally. Maize serves as a modest source of dietary fat, contributing 1.4 g of fat per day, which represents 1.6% of daily intake. However, its role in nutrition is considerably enhanced through its indirect contribution as livestock feed.

Maize in Livestock Feed:

- Over 50% of global maize production is used as feed for monogastric (e.g., poultry) and ruminant (e.g., cattle) livestock.
- Industrial maize processing (e.g., wet-milling, dry-milling, and distilling) produces valuable by-products that provide additional nutrients for livestock (Loy & Lundy, 2019).

- Bioethanol production from maize generates distiller's dry grains, a high-value feed co-product that can substitute 1.2 tons of maize grain with 1 ton of distiller's dry grains (Wallington et al., 2012).

Specialized Maize for Livestock:

- In high-income countries, maize varieties with improved nutritional value are being developed specifically for livestock feed (Loy & Lundy, 2019).
- Feed applications are becoming more specialized, targeting poultry and livestock nutritional needs (Paulsen et al., 2019).

Maize is widely cultivated for forage, covering an estimated 16.8 M ha annually (Heuzé et al., 2017b). Maize silage has become a dominant feed source for livestock in temperate regions, often surpassing grain cultivation in cooler areas such as the United Kingdom. In many parts of the Global South, maize stover (the remaining plant material after maize grain harvest) serves as a critical livestock forage resource (Valbuena et al., 2012; Tittonell et al., 2015). Research is increasingly focused on developing dual-purpose maize varieties that provide both grain for human consumption and stover for livestock feed (Blümmel et al., 2013). This approach holds particular value for resource-constrained regions, improving food security and livestock nutrition simultaneously.

2.6 MAIZE CULTIVATION: INDIAN PERSPECTIVE

2.6.1 Climate and Soil Requirements

Maize is predominantly a warm-weather crop that thrives across diverse climatic conditions. It can successfully grow in regions receiving an annual rainfall of 60 cm, provided it is well-distributed throughout its growth stages. The crop requires over 50% of its total water needs between 30 and 35 days after tasseling, and inadequate moisture during the grain filling stage results in poor yields and shriveled grains. Maize is highly susceptible to frost damage at all growth stages. While prolonged cloudy weather is harmful, a combination of intermittent sunlight and rainfall creates ideal conditions for growth. Bright, sunny days promote photosynthesis and accelerate plant growth. In India, maize is traditionally cultivated during the monsoon (Kharif) season, characterized by high temperatures (up to 35°C) and frequent rains. However, with the development of improved cultivars and modern production technologies, winter maize cultivation has emerged as a viable alternative. Ideal soil types for maize include loam or silt loam surface soils with permeable subsoil, as well as brown silt clay loam with moderate permeability. Deep, fertile soils enriched with organic matter and well-drained conditions are optimal for maize growth. A soil pH between 7.5 and 8.5 is favorable, as values beyond this range may result in nutrient imbalances and elemental toxicity. Since 85% of maize in India is grown during the monsoon season, effective soil water retention and drainage are crucial to prevent waterlogging and reduce the risk of seed and seedling diseases (Tripathi et al., 2011). Meeting these environmental and soil requirements is essential for achieving improved productivity and better resilience in maize cultivation.

2.6.2 Zonalization of Varietal Testing System

The Directorate of Maize Research (DMR), which was upgraded in January 1994 from its previous status as the All India Coordinated Maize Improvement Project (AICMIP), is responsible for conducting and coordinating maize research, developing improved technologies to enhance maize productivity, and promoting the diversified uses of maize products (Tripathi et al., 2011). For effective varietal evaluation and development, the DMR has categorized India into five distinct agro-climatic zones to facilitate region-specific testing and adaptation of maize varieties.

i. Zone 1—Northern India Hills Zone: This is further subdivided into 1-A comprising Jammu & Kashmir, Himachal Pradesh, and Uttaranchal. 1-B comprising North-eastern India and Assam.
ii. Zone 2—Northern Plains Zone: This comprises Punjab, Haryana, Delhi, and western Uttar Pradesh.
iii. Zone 3—Comprising Rajasthan, Gujarat, and Madhya Pradesh
iv. Zone 4—Comprising Peninsular India.
v. Zone 5—New Maize-growing areas like Orissa, Jharkhand, and West Bengal.

2.6.3 Status of Maize Cultivation

Maize holds a significant position in Indian agriculture, ranking as the third most important cereal crop after wheat and rice. Its versatility, adaptability to diverse agro-climatic conditions, and wide range of uses contribute to its prominence in the country's cropping systems. Maize is cultivated extensively across various regions in India, playing a crucial role in food security, livestock feed, and industrial applications. The major maize-growing states are Uttar Pradesh, Bihar, Rajasthan, Madhya Pradesh, Punjab, Andhra Pradesh, Himachal Pradesh, West Bengal, Karnataka, and Jammu & Kashmir, jointly accounting for over 95% of the national maize production. The estimate of maize production in 2007 was 13–14 M t from an area of 7.2 M ha with an average productivity of 2 tons/hectare.

Maize is traditionally cultivated during the summer (monsoon) season in India, characterized by high temperatures (<35°C) and rainfall. However, Rabi (winter) maize cultivation, introduced in the mid-sixties in select regions of Bihar and South India, has now expanded nationwide. Rabi maize offers several advantages, including reduced incidence of diseases and insect pests, minimal damage from heavy rainfall, and slower weed growth, making it a preferred choice for many farmers (Tripathi et al., 2011).

Although the area under maize cultivation has shown an increasing trend, achieving substantially higher production is crucial to meet the rising demand for maize in human food, animal feed, poultry feed, and industrial processing for value-added products. Biotechnology has emerged as a significant technique for enhancing maize productivity by improving resistance to major insect pests and developing herbicide-tolerant varieties (James, 2008). Additionally, traits such as high lysine content, amylase enzyme production, phytase enzyme expression, and drought tolerance are being incorporated into maize (Stein & Rodriguez-Cerezo, 2009). Globally, several successful biotech interventions have been applied to maize, and in India, confined field trials of transgenic maize with insect resistance and herbicide-tolerant traits are currently underway.

2.7 BIOCHEMISTRY OF MAIZE KERNEL

Maize kernels, similar to those in other cereal crops, serve as storage organs that contain vital components essential for plant growth and reproduction (Table 2.4).

The kernel is composed of three distinct compartments that contribute to its overall energy density: the persistent endosperm (83%), a sizable embryo (11%), and a maternal pericarp (6%). Maize kernels are rich in nutrients, containing approximately 72% starch, 10% protein, 4% lipid, and various micronutrients such as vitamins and minerals (Nuss & Tanumihardjo, 2010) (Table 2.5).

TABLE 2.4
Relative Content (%) of Nutrient Components in Maize Kernel

	Pericarp (Seed Coat)	Endosperm	Embryo
Protein	3.7	8.0	18.4
Crude fiber	86.7	2.7	8.8
Crude fat	1.0	0.8	33.2
Starch	7.3	87.6	8.3
Sugar	0.34	0.62	10.8

TABLE 2.5
Nutritional Composition of Edible Portion of Maize

Calories	342.0	***Calcium***	10.0
Moisture (g)	14.9	***Iron***	2.3
Carbohydrates (g)	66.2	***Potassium***	286.0
Protein	11.1	***Magnesium***	139.0
Fat	3.6	***Copper***	0.14
Fiber	2.7	***Amino acids***	1.78
Minerals	1.5	***Riboflavin***	0.10
Phosphorus	348.0	***Thiamine***	0.42
Sodium	15.9	***Vitamin C***	0.12
Sulphur	114.0	***Carotene***	90.0

Source: FAOSTAT Database, 2012.

Notably, maize contributes around 15% of the world's protein and 20% of the global calorie supply, underscoring its significant role in global nutrition (Council, 1988; Shiferaw et al., 2011).

2.7.1 Starch

Maize is the primary raw material for the starch industry, accounting for approximately 90% of global starch production (Tetlow, 2011). Starch synthesis, extensively reviewed in multiple studies (Wilson et al., 2004; Dai et al., 2021; Qu et al., 2022), occurs within the amyloplasts of maize endosperm cells, forming starch granules as part of a mega-molecular assembly. These granules contain two major glucose homopolymers: amylose and amylopectin. Amylose is composed predominantly of linear glucose chains linked by α-1,4 glycosidic bonds. In contrast, amylopectin features a branched structure with α-1,4 linkages as the primary connections, while α-1,6 linkages create branch points. The alternating regions of branching and non-branching enable the alignment of linear glucose chains in parallel helices, giving starch its semi-crystalline structure. The carbon skeletons required for starch synthesis are derived from sucrose transported to the endosperm. This process involves specific enzyme isoforms unique to cereal endosperms. As endosperm cells mature, starch granules expand, with the amylose content increasing relative to amylopectin. Typically, mature starch granules maintain a 3:1 ratio of amylopectin to amylose. Within each endosperm cell type, the starch granules are uniform in size. A single mature granule may contain approximately 10^9 amylose and 10^7 amylopectin molecules, which are systematically arranged to provide stability and crystalline properties. Normal maize starch consists of amylose and amylopectin, with its biosynthesis regulated by four key enzymes: ADP-glucose pyrophosphorylase (AGPase), starch synthase (SS), starch branching enzyme (SBE), and starch debranching enzyme (Li et al., 2018). Mutations in genes associated with starch biosynthesis can alter starch development and deposition, resulting in structural modifications that ultimately produce starch with distinct functional properties (Tziotis et al., 2005; Li et al., 2010; Zhang et al., 2019).

2.7.2 Protein

Discussion of maize kernel proteins has been extensively covered in several studies (Boyer & Hannah, 1994; Monjardino et al., 2005; Monjardino et al., 2006). Maize kernels are composed of approximately 10% protein, 72% starch, and 4% fat (Peña-Rosas et al., 2014). The storage protein, a 7S globulin, is present in both the embryo and the endosperm. While the embryo has a higher protein concentration, the endosperm contributes the greatest total protein content due to its larger proportion within the kernel (FAO, 1992). Endosperm proteins are classified into prolamins, collectively known as zeins (approximately 52% of kernel nitrogen), glutelins (about 25%), albumins (around 7%), and globulins (around 5%). Zeins, which accumulate in the rough endoplasmic reticulum of endosperm cells, are rich in the amino acids glutamine, proline, leucine, and alanine but are notably deficient in the essential amino acids lysine and tryptophan. In contrast, embryo proteins provide higher concentrations of tryptophan and lysine (FAO, 1992). Zeins have been identified through

SDS-PAGE analysis and are classified based on their solubility and structural properties into four groups: α- (22 and 19 kD), β- (15 kD), γ- (27, 16, and 50 kD), and δ- (10 and 18 kD) zeins. Among these, α-zeins are the most abundant, constituting approximately 80% of total zein content, and are encoded by a large family of genes, whereas the other zein classes are encoded by only one or two genes each. Since the endosperm forms the bulk of maize kernel weight, its protein composition significantly influences kernel texture and nutritional quality. Zeins account for over half of the total endosperm protein, and changes in their content can markedly impact both the nutritional value and visual characteristics of maize kernels.

Traditional maize endosperm protein is nutritionally inadequate due to its low content of essential amino acids such as lysine (0.150–0.250% in flour) and tryptophan (0.030–0.040% in flour) (Hossain et al., 2018). Since these amino acids are not synthesized by monogastric animals, including humans, they must be obtained through dietary sources (Prasanna et al., 2020). The presence of the mutant *o2* gene improves protein quality by increasing lysine and tryptophan levels, while the recessive *wx1* gene promotes the accumulation of easily digestible amylopectin, enhancing maize's potential as a rapid energy source. Both genetic traits are valuable in addressing malnutrition (Talukder et al., 2022a). Several classical *opaque* mutants, characterized by their distinct maize kernel phenotype, have been identified; these mutants often result from reduced zein synthesis in the endosperm. Some mutants display abnormal protein body formation, which is crucial for proper zein accumulation. Additionally, RNA interference-mediated inhibition of zein gene expression can mimic the *opaque* phenotype (Wu & Messing, 2010). However, in the classic maize mutant *opaque1* (O1), the quantity and amino acid profile of zein and non-zein proteins are comparable to the wild type, suggesting no apparent defect in zein synthesis in this variant (Bertonii, 2012).

In addition to their desirable high lysine and tryptophan content, these mutants exhibit agronomically undesirable traits such as reduced yield and a soft, floury endosperm, which increases susceptibility to diseases and insect attacks while also complicating food processing (Ufaz & Galili, 2008). To overcome these limitations, breeding efforts at the International Maize and Wheat Improvement Center (CIMMYT) in Mexico during the early 1990s led to the development of *opaque2*-derived QPM lines, which combine improved agronomic performance with enhanced nutritional value. These QPM varieties have since been successfully adopted in several countries (Ufaz & Galili, 2008). Moreover, QPM varieties with elevated tryptophan content may help mitigate niacin deficiency, a common nutritional concern in maize-based diets.

2.7.3 Lipids

Lipids (oil) in maize are primarily concentrated in the embryo, particularly in the scutellum, where they constitute about 40% of its dry weight. These lipids serve as a crucial energy source through gluconeogenesis to support the developing embryo after germination (Oaks & Beevers, 1964). The embryo itself contains roughly 33% oil, while the total oil content in a typical maize kernel is approximately 4% (The oil quantity and composition in maize kernels are genetically regulated, and through intensive selection efforts, such as in the Illinois High Oil line, the kernel oil content has been progressively increased to over 20% after 106 generations spanning more than a century (Dudley, 2007).

Fatty acids in maize oil are predominantly found in the form of triacylglycerides, where they are esterified to the hydroxyl groups of glycerol. These triacylglycerides comprise a combination of saturated and unsaturated fatty acids with the following approximate distribution (Boyer & Hannah, 1994):

- **Saturated fatty acids:** 12% palmitic acid (16:0) and 2% stearic acid (18:0).
- **Unsaturated fatty acids:** 25% oleic acid (18:1), 60% linoleic acid (18:2), and 1% linolenic acid (18:3).

Maize oil is widely recognized for its high quality, attributed to its elevated linoleic acid content and minimal linolenic acid levels. However, some studies suggest that the breakdown of esterified linolenic acid into its free form during maize meal storage may be linked to an increased risk of esophageal carcinogenesis (Sammon & Iputo, 2006). The genetic regulation of maize oil lipid content is complex, likely governed by the inheritance of multiple genes, reflecting a polygenic control mechanism (Boyer & Hannah, 1994).

2.7.4 Toxins

Feeding livestock with maize, particularly in its green chop form or when animals graze on maize fields, can sometimes lead to toxic reactions caused by nitrate or nitrite poisoning. This occurs when bacteria associated with maize plants convert nitrates into toxic nitrites or nitrogen dioxide, posing risks to both animals and humans. Additionally, maize plants and grains are susceptible to fungal infections, which can lead to the production of harmful mycotoxins that further contribute to toxicity concerns.

2.7.4.1 Nitrate Poisoning

Nitrate is naturally absorbed by plants and utilized in the biosynthesis of proteins and other nitrogen-containing compounds (Pavelchak, 1999). However, under certain environmental conditions, such as after a drought ends, some plants may accumulate nitrate at elevated levels if it is abundantly present in the soil. Maize, particularly in its lower stalk regions (below 20 cm), can accumulate nitrate concentrations that may become toxic to livestock (Everist, 1981). Ruminants are generally less vulnerable to nitrate poisoning if their diet includes sufficient carbohydrates. This is because rumen microorganisms can convert nitrate to nitrite, which is subsequently transformed into ammonia by other microbial species within the rumen. Nevertheless, excessive nitrate intake can overwhelm this process, leading to toxicity. Symptoms of nitrate poisoning include rapid breathing, a fast yet weak heartbeat, difficulty in breathing, muscle tremors, staggering, and in severe cases, death

2.7.5 Allergens

Although maize is not typically considered a major allergenic crop, some allergic reactions have been documented in scientific literature. Maize can induce both food and inhalation allergies, with pollen and maize dust being known triggers for conditions such as hay fever and baker's asthma.

The International Union of Immunological Societies identifies four recognized maize allergens: Zea m 1, Zea m 12, Zea m 14, and Zea m 25. Among these, **Zea m 1** is a β-expansin protein (EXPB) with a molecular weight of approximately 27 kDa, closely related to the **Lol p 1** allergen found in ryegrass (*Lolium perenne*) (Broadwater et al., 1993). Zea m 1 exists in four isoforms in maize pollen, with **EXPB1** being the most abundant (Yennawar et al., 2006).

Zea m 12, a profilin with a molecular weight of approximately 14 kDa, is encoded by a multi-gene family in maize (Staiger et al., 1993). Profilins are common allergens in various plant species, including alder (*Alnus glutinosa*), hazel (*Corylus avellana*), timothy grass (*Phleum pratense*), and rye (*Secale cereale*). Notably, a birch (*Betula verrucosa*) profilin was identified as a major pollen allergen in a study by Valenta et al. (1991).

In a study conducted by Pastorello et al. (2000), a major maize food allergen was identified as a 9 kDa lipid transfer protein (LTP), known as Zea m 14 (Swiss Prot Data Bank, accession number P19656). This LTP shares homology with LTPs from rice, peach, and apricot. Cross-reactivity was observed, as pre-incubation of immune sera from maize-allergic patients with rice or peach extracts inhibited antibody binding in immunoblots, confirming their antigenic similarity. The second significant food allergen identified was a 16 kDa trypsin/activated Hageman factor (XIIA) inhibitor (Swiss Prot Data Bank, accession number P01088). Similar inhibitors found in other cereals have been linked to Baker's asthma, a respiratory condition common among bakers. Immune sera from allergic patients, when pre-incubated with Timothy grass pollen, wheat, rice, or barley extract, completely inhibited antibody binding to this 16 kDa allergen and other minor maize seed allergens, further indicating cross-reactivity. In addition to the previously described allergens like β-expansins, profilins, and LTPs, Weichel et al. (2006) identified thioredoxins, such as Zea m 25, in enriched repertoires of maize IgE-binding sequences, expanding the list of recognized maize allergens.

Pasini et al. (2002) identified a 50 kDa protein in maize as a major allergen that exhibits notable heat stability and resistance to peptic and pancreatic digestion, making it particularly persistent in the digestive tract. In a separate study, Lee et al. (2005) described a 50 kDa γ-zein from maize that demonstrated strong in vitro cross-reactivity with the almond major protein (AMP), a key allergen in almonds (*Prunus amygdalus*). However, it remains uncertain whether these two identified 50 kDa proteins correspond to the same maize allergen.

2.7.6 Other Undesirable Phytochemicals

Maize contains low levels of certain anti-nutrients, which hinder the normal absorption and utilization of nutrients. One such compound is phytic acid (or phytate), the primary storage form of phosphorus in plants. Phytate is poorly accessible to humans and non-ruminant animals because they lack the enzyme phytase required to break it down. This poses several challenges:

- **Nutritional Impact:** Humans may need to obtain phosphorus from alternative dietary sources.

- **Animal Feed Efficiency:** Livestock feed often requires phosphorus supplementation to compensate for phytate-bound phosphorus.
- **Environmental Concerns:** The excess phosphorus from fortified feeds is excreted, contributing to environmental pollution.

Additionally, phytate acts as an anti-nutrient by chelating essential metal ions such as iron, reducing their bioavailability in the diet (Hurrell, 2003). Recent advancements in genetic engineering have led to maize varieties that express phytase, an enzyme that enhances phosphorus bioavailability, thereby improving nutritional outcomes and reducing environmental concerns (Drakakaki et al., 2005; Chen et al., 2007). Another anti-nutrient that is present in maize is the sugar raffinose. Raffinose is non-digestible, and is considered an anti-nutrient due to gas production and resulting flatulence. It can be removed from food and feed by soaking, cooking, and irradiation or by enzyme or solvent treatment (OECD, 2002). Both trypsin and chymotrypsin inhibitors are present at low levels in maize (OECD, 2002). They are not considered important for human or animal nutrition.

2.7.7 Beneficial Phytochemicals

Maize is recognized as a vital food crop for humans and a high-energy feed for animals (Tanumihardjo et al., 2020). In the human diet, it serves as a valuable source of vitamin B1, vitamin B5, folate, dietary fiber, vitamin C, phosphorus, and manganese. As a staple, maize compares favorably with root and tuber crops and provides an energy value similar to dried legumes (Okoruwa & Kling, 1996). However, processing tends to reduce the concentrations of proteins, lipids, and fiber, potentially lowering its overall nutritional profile (FAO, 1992).

2.8 MAIZE IN PHARMACOLOGY

Maize has been traditionally regarded as a remedy for various health conditions, including emaciation, eating disorders, and hemorrhoids. It is known to possess analgesic properties and is believed to enhance sexual well-being, particularly in men with erectile dysfunction, by stimulating the production of sex-related hormones. Maize silk has been widely used in folk medicine across multiple cultures to treat urinary tract infections and kidney stones. In China, maize silk is recognized for its role in alleviating fluid retention in the liver, jaundice, and for promoting liver function and bile production. Additionally, maize silk is used to lower blood pressure and support urinary health. A decoction made from maize leaves, roots, and silk is traditionally used to manage bladder symptoms, while a cob tea decoction is employed for gut ailments. Maize silk also acts as an emollient for treating wounds, ulcers, and edema (Kumar & Jhariya 2013).

Maize-derived phenolic acids have demonstrated anti-diabetic properties by improving glucose uptake, promoting glycogen synthesis, and improving lipid and glucose profiles in relation to obesity and cardiovascular diseases (Vinayagam et al., 2016). They reduce blood glucose levels, inhibit glucose-6-phosphatase and phosphoenolpyruvate carboxykinase activity, decrease lipid peroxidation, and increase both glycogen and insulin levels (Senaphan et al., 2015). Maize peptides exhibit a range of biological effects, including antioxidant, anti-cancer, anti-inflammatory, hepatoprotective, alcohol metabolism facilitation, antimicrobial, and anti-hypertensive properties. They have also been found to inhibit dipeptidyl peptidase IV, contributing to improved glycemic control (Trinidad-Calderon et al., 2021).

Anthocyanins found in maize offer strong protection against oxidative stress, contributing to the management of chronic diseases (MANN 2014). Additionally, maize silk's mild diuretic effect, coupled with its demulcent properties, provides relief from cystitis, bladder irritation, and uric acid and phosphate stones. Chinese medicine also utilizes maize silk for treating edema. Phytochemical studies have identified beneficial compounds in maize silk, including chlorogenic acid, ferulic acid, saponins, phytosterols, resin, and sugars (Van Damme, 2022).

REFERENCES

Afidchao, M. M., Musters, C. J. M., Wossink, A., Balderama, O. F., & De Snoo, G. R. (2014). Analysing the farm level economic impact of GM corn in the Philippines. *NJAS - Wageningen Journal of Life Sciences*, 70, 113–121.

Aguirre, G., J. A., Bellon, M. R., & Smale, M. 1998. A regional analysis of maize biological diversity in southeastern Guanajuato, Mexico. CIMMYT economics Working Paper 98-06. Mexico, D. F.: CIMMYT.

Ahloowalia, B. S., & Dhawan, N. L. (1963). Effect of genetic diversity on combining ability of inbreds of maize. *Indian Journal of Genetics*, 23, 158–162.

Alotaibi, M. (2023). Climate change, its impact on crop production, challenges, and possible solutions. *Notulae Botanicae Horti Agrobotanici Cluj-Napoca*, 51, 13020. doi: 10.15835/nbha51113020

Azadi, M. S., Shokoohfar, A., Mojdam, M., Lak, S., & Fazel, M. A. (2022). Investigation of changes in corn hybrids grain protein, proline, and micronutrient content under the influence of drought and fertilizers. *Turkish Journal of Agriculture and Forestry*, 46(4), 413–423.

Balmer, D., Flors, V., Glauser, G., & Mauch-Mani, B. (2013). Metabolomics of cereals under biotic stress: Current knowledge and techniques. *Frontiers in Plant Science*, 4, 82. doi: 10.3389/fpls.2013.00082

Barea, J. M. (2015). Future challenges and perspectives for applying microbial biotechnology in sustainable agriculture based on a better understanding of plantmicrobiome interactions. *Journal of Soil Science and Plant Nutrition*, 15, 261–282. doi: 10.4067/S0718-95162015005000021

Beckett, J. B. (1971). Classification of male-sterile cytoplasms in maize (Zea mays L.). *Crop Science*, 11(5), 724–727.

Bellon, M. R., Hodson, D., Bergvinson, D., Beck, D., Martinez-Romero, E., & Montoya, Y. (2005). Targeting agricultural research to benefit poor farmers: Relating poverty mapping to maize environments in Mexico. *Food Policy*, 30, 476–492.

Berta, G., Copetta, A., Gamalero, E., Bona, E., Cesaro, P., & Scarafoni, A. et al. (2014). Maize development and grain quality are differentially affected by mycorrhizal fungi and a growth-promoting pseudomonad in the field. *Mycorrhiza*, 24, 161–170. doi: 10.1007/s00572-013-0523-x

Bertoni, G. (2012). Maize *opaque1* and protein body formation. *The Plant Cell*, 24(8), 3168. doi: 10.1105/tpc.112.240811

Betrán, F. J., Beck, D., Bänziger, M., & Edmeades, G. O. (2003). Secondary traits in parental inbreds and hybrids under stress and non-stress environments in tropical maize. *Field Crops Research*, 83(1), 51–65.

Bhusal, B., Poudel, M. R., Rishav, P., Regmi, R., Neupane, P., & Bhattarai, K. et al. (2021). A review on abiotic stress resistance in maize (Zea mays L.): Effects, resistance mechanisms and management. *Journal of Biology and Today's World*, 10, 1–3.

Blare, T., Donovan, J., & Garcia-Medina, M. (2020). The right tortilla for the right occasion: Variation in consumers' willingness to pay for blue maize tortillas based on utilization. *Journal of Food Products Marketing*, 26, 564–579.

Blümmel, M., Grings, E., & Erenstein, O. (2013). Potential for dualpurpose maize varieties to meet changing maize demands: Synthesis. *Field Crops Research*, 153, 107–112.

Boddupalli, P., Suresh, L. M., Mwatuni, F., Beyene, Y., Makumbi, D., Gowda, M., ... & Lassen, P. (2020). Maize lethal necrosis (MLN): Efforts toward containing the spread and impact of a devastating transboundary disease in sub-Saharan Africa. *Virus Research*, 282, 197943.

Boyer, C. D., & Hannah, L. C. (1994). Kernel mutants of corn. Chapter 1. In: A. R. Hallauer, ed. *Specialty corns*. Boca Raton, USA: CRC Press Inc., 1–28.

Broadwater, A. H., Rubinstein, A. L., Chay, C. H., Klapper, D. G., & Bedinger, P. A. (1993). *Zea ml,* the maize homolog of the allergen-encoding *Lol pl* gene of rye grass. *Gene*, 131, 227–230.

Brookes, G., & Barfoot, P. (2020). GM crop technology use 1996–2018: Farm income and production impacts. *GM Crops & Food*, 11, 242–261.

Bruce, W. B., Edmeades, G. O., & Barker, T. C. (2002). Molecular and physiological approaches to maize improvement for drought tolerance. *Journal of Experimental Botany*, 53(366), 13–25.

Burdon, J. J., & Zhan, J. (2020). Climate change and disease in plant communities. *PLoS Biology*, 18, e3000949.

Burlakoti, S., Devkota, A. R., Poudyal, S., & Kaundal, A. (2024). Beneficial plant– microbe interactions and stress tolerance in maize. *Applied Microbiology*, 4, 1000–1015. doi: 10.3390/applmicrobiol4030068

Cairns, J., Hellin, J., Sonder, K., Araus, J., MacRobert, J., Thierfelder, C., & Prasanna, B. M. (2013). Adapting maize production to climate change in sub-Saharan Africa. *Food Security*, 5, 345–360.

Chen, R., Xue, G., Chen, P., Yao, B., Yang, W., Ma, Q., Fan, Y., Zhao, Z., Tarczynski, M., & Shi, J. (2007). Transgenic maize plants expressing a fungal phytase gene. *Transgenic Research*, 17, 633–643.

Coghlan, A. (2000). For the moment, the gene genie is staying in its bottle. *New Scientist*, 25.

Dai, D., Ma, Z., & Song, R. (2021). Maize kernel development. *Molecular Breeding*, 41, 1–33.

Dartora, J., Guimarães, V. F., Menezes, C. R., Freiberger, M. B., Castoldi, G., & Gonçalves, E. (2016). Maize response to inoculation with strains of plant growth promoting bacteria. *Revista Brasileira de Engenharia Agrícola e Ambiental,* 20, 606–611. doi: 10.1590/1807-1929/agriambi.v20n7p606-611

De Groote, H., Kimenju, S. C., Munyua, B., Palmas, S., Kassie, M., & Bruce, A. (2020). Spread and impact of fall armyworm (Spodoptera frugiperda J.E. Smith) in maize production areas of Kenya. *Agriculture, Ecosystems & Environment*, 292, 106804.

Deutsch, C. A., Tewksbury, J. J., Tigchelaar, M., Battisti, D. S., Merrill, S. C., Huey, R. B., & Naylor, R. L. (2018). Increase in crop losses to insect pests in a warming climate. *Science*, 361, 916–919.

De Vries, B. D., Shuler, S. L., & Tracy, W. F. (2016). Endosperm carbohydrates in pseudostarchy and extreme-sugary maize inbreds during kernel development. *Crop Science*, 56(5), 2448–2456.

Dhawan, N. L., & Singh, J. (1961). Flint x Dent maize hybrids give increased yields. *Current Science*, 30, 233–234.

Dhawi, F., Datta, R., & Ramakrishna, W. (2015). Mycorrhiza and PGPB modulate maize biomass, nutrient uptake and metabolic pathways in maize grown in mining impacted soil. *Plant Physiology and Biochemistry*, 97, 390–399. doi: 10.1016/j.plaphy.2015.10.028

Di-Giovanni, Kevan, F. P. G., & Nasr, M. E. (1995). The variability in settling velocities of some pollen and spores. *Grana*, 34, 39–44.
Dodd, I. C., & Ruiz-Lozano, J. M. (2012). Microbial enhancement of crop resource use efficiency. *Current Opinion in Biotechnology*, 23, 236–242. doi: 10.1016/j.copbio.2011.09.005
Doebley, J. (1990). Molecular evidence for gene flow among *Zea* species. *BioScience*, 40(6), 443–448.
Drakakaki, G., Marcel, S., Glahn, R., Lund, E., Pariagh, S., Fischer, R., Christou, P., & Stoger, E. (2005). Endosperm-specific co-expression of recombinant soybean ferritin and *Aspergillus* phytase in maize results in significant increases in the levels of bioavailable iron. *Plant Molecular Biology*, 59, 869–880.
Dudley, J. W. (2007). From means to QTL: The Illinois long-term selection experiment as a case study in quantitative genetics. *Crop Science*, 47(S3), S20–S31.
Duvick, D. N. (2005). The contribution of breeding to yield advances in maize (Zea mays L.). *Advances in Agronomy*, 86, 83–145. doi: 10.1016/S0065-2113(05)86002-X
Du Plessis, D. P., & Dijkhuis, F. J. (1967). The influence of the time lag between pollen-shedding and silking on the yield of maize.
Edmeades, G. O., Bolaños, J., Chapman, S. C., Lafitte, H. R., & Bänziger, M. (1999). Selection improves drought tolerance in tropical maize populations: I. Gains in biomass, grain yield, and harvest index. *Crop Science*, 39(5), 1306–1315.
Erenstein, O., Chamberlin, J., & Sonder, K. (2021). Estimating the global number and distribution of maize and wheat farms. *Global Food Security*, 30, 100558.
Eubanks, M. W. (1993). A cross between *Tripsacum dactyloides* and *Zea diploperennis. Maize Newsletter*, 67, 39.
Eubanks, M. W. (1995). A cross between two maize relatives: *Tripascum dactyloides* and *Zea diploperennis* (Poaceae). *Economic Botany*, 49(2), 172–182.
Eubanks, M. W. (1997a). Further evidence for two progenitors in the origin of maize. *Maize Newsletter*, 71, 35–36.
Eubanks, M. W. (1997b). Molecular analysis of crosses between *Tripsacum dactyloides* and *Zea diploperennis* (Poaceae). *Theoretical and Applied Genetics*, 94, 707–712.
Everist, S. L. (1981). *Poisonous plants of Australia*, Sydney, Angus & Robertson Publishers.
FAO & CIMMYT (1997). *White maize: A traditional food grain in developing countries*, Rome, FAO & CIMMYT.
FAO (1992). Maize in Human Nutrition. Food and Agriculture Organization of the United Nations, available online at http://www.fao.org/docrep/t0395e/T0395E00.htm#Contents Rome.
FAOSTAT (2012) Food and Agriculture Organization of the United Nations Data. Statistical Databases. http://faostat.fao.org/default.aspx?lang=en
FAOStat. (2021). FAO Stat. FAO, Rome. http://www.fao.org/faostat
Frey, T. J., Weldekidan, T., Colbert, T., Wolters, P. J. C. C., & Hawk, J. A. (2011). Fitness evaluation of Rcg1, a locus that confers resistance to Colletotrichum graminicola (Ces.) GW Wils. using near-isogenic maize hybrids. *Crop Science*, 51, 1551–1563. doi: 10.2135/cropsci2010.10.0613
García-Lara, S., Chuck-Hernandez, C., & Serna-Saldivar, S. O. (2019). Development and structure of the corn kernel. In: Serna-Saldivar, S.O., ed. *Corn* (Third Edition). Oxford: AACC International Press.

Genetics

Gould, F. W. (1968). *Grass systematics*, New York, USA, McGraw Hill.
Hellin, J., Dixon, J., Higman, S., & Keleman, A. (2011). High-value agricultural products and poverty reduction: Smallholder farmer access to maize markets. *Journal of Crop Improvement*, 25, 371–391.

Heuzé, V., Tran, G., Edouard, N., & Lebas, F. (2017b). Maize silage. INRAE, CIRAD, AFZ and FAO.

Hitchcock, A. S., & Chase, A. 1951. *Manual of the grasses of the United States*. Vol. 2, New York, Dover Publications, 790–796.

Hossain, F., Muthusamy, V., Pandey, N., Vishwakarma, A. K., Baveja, A., Zunjare, R. U., & Gupta, H. S. (2018). Marker-assisted introgression of opaque2 allele for rapid conversion of elite hybrids into quality protein maize. *Journal of Genetics*, 97, 287–298. https://www.ontario.ca/page/potential-nitrate-poisoning-and-silo-gas-when-using-dry-weather-damaged-corn updated June 18, 2024.

Hudson, A. I., Odell, S. G., Dubreuil, P., Tixier, M. H., Praud, S., & Runcie, D. E. et al. (2022). Analysis of genotype-by-environment interactions in a maize mapping population. *G3 (Bethesda)*, 12, jkac013. doi: 10.1093/g3journal/jkac013

Hurrell, R. F. (2003). Influence of vegetable protein sources on trace element and mineral bioavailability. *Journal of Nutrition*, 133, 2973S–2977S.

ISAAA. (2019). Global Status of Commercialized Biotech/GM Crops in 2019: Biotech Crops Drive Socio-Economic Development and Sustainable Environment in the New Frontier. ISAAA Briefs 55. ISAAA, Ithaca, NY. http://www.isaaa.org/resources/publications/briefs/

James, C. 2008. Global status of commercialized biotech/GM crops: 2008, Brief No. 39, International Services for the Acquisition of Agribiotech Applications (ISAAA).

Jones, P. G., & Thornton, P. K. (2003). The potential impacts of climate change on maize production in Africa and Latin America in 2055. *Global Environmental Change*, 13, 51–59.

Kassie, M., Wossen, T., De Groote, H., Tefera, T., Sevgan, S., & Balew, S. (2020). Economic impacts of fall armyworm and its management strategies: Evidence from southern Ethiopia. *European Review of Agricultural Economics*, 47, 1473–1501.

Katiyar, S. K., & Sachan, J. K. S. (1992). Isozyme diversity in *Zea* and related genera. *Maize Newsletter*, 66, 92.

Katiyar, S. K., Chandel, G., Kumar, R., & Sachan, J. K. S. (1992). Karyological studies in *Coix* and maize. *Maize Newsletter*, 66, 90.

Keleman, A., & Hellin, J. (2009). Specialty maize varieties in Mexico: A case study in market-driven agro-biodiversity conservation. *Journal of Latin American Geography*, 8, 147–174.

Khosravi, A. R., Mansouri, M., Bahonar, A. R., & Shokri, H. (2007). Mycoflora of maize harvested from Iran and imported maize. *Pakistan Journal of Biological Sciences: PJBS*, 10(24), 4432–4437.10.

Kimotho, R. N., Baillo, E. H., & Zhang, Z. (2019). Transcription factors involved in abiotic stress responses in Maize (Zea mays L.) and their roles in enhanced productivity in the post genomics era. *PeerJ*, 7, e7211. doi: 10.7717/peerj.7211

Kloepper, J. W. (1978). Plant growth-promoting rhizobacteria on radishes, in Proceedings of the 4th International Conference on Plant Pathogenic Bacteria, Angers, France, 1978, Vol. 2, 879–882 (INRA, Angers, France: Station de Pathologie Vegetale et Phytobacteriologie, INRA).

Kumar, D., & Jhariya, A. N. (2013). Nutritional, medicinal and economical importance of corn: A mini review. *Journal of Pharmaceutical Sciences and Research*, 2319, 555X.

Kumar, D., & Singh, V. (2019). Bioethanol production from corn. In: Serna-Saldivar, S. O., ed. *Corn* (Third Edition). Oxford: AACC International Press, 615–631.

Laurie, D. A., & Bennett, M. D. (1986). Wheat x maize hybridization. *Canadian Journal of Genetics and Cytology*, 28, 313–316.

Lee, S. H., Benmoussa, M., Sathe, S. K., Roux, K. H., Teuber, S. S., & Hamaker, B. R. (2005). A 50 kDa maize γ-zein has marked cross-reactivity with the almond major protein. *Journal of Agricultural and Food Chemistry*, 53, 7965–7970.

Li, X., Chen, G. H., Zhang, W. Y., & Zhang, X. (2010). Genome-wide transcriptional analysis of maize endosperm in response to ae wx double mutations. *Journal of Genetics and Genomics*, 37(11), 749–762.

Li, X.-J., Wu, M.-F., Ma, J., Gao, B.-Y., Wu, Q.-L., Chen, A.-D., Liu, J., Jiang, Y.-Y., Zhai, B.-P., Early, R., Chapman, J. W., & Hu, G. (2020). Prediction of migratory routes of the invasive fall armyworm in eastern China using a trajectory analytical approach. *Pest Management Science*, 76, 454–463.

Li, Y. P., Yu, G. W., Lv, Y. N., Long, T. D., Li, P., Hu, Y. F., Liu, H. M., Zhang, J. J., Liu, Y. H., Li, W. C., & Huang, Y. (2018). Combinatorial interaction of two adjacent cis-active promoter regions mediates the synergistic induction of Bt2 gene by sucrose and ABA in maize endosperm. *Plant Science*, 274, 332–340.

Lobell, D. B., Roberts, M. J., Schlenker, W., Braun, N., Little, B. B., & Rejesus, R. M. et al. (2014). Greater sensitivity to drought accompanies maize yield increase in the US Midwest. *Science*, 344, 516–519. doi: 10.1126/science.1251423

Lodha, T. D., Hembram, P., & Nitile Tep, J. B. (2013). Proteomics: A successful approach to understand the molecular mechanism of plant-pathogen interaction. *American Journal of Plant Sciences*, 4(6), 15. doi: 10.4236/ajps.2013.46149

Louette, D., & Smale, M. 1998. Farmers' Seed Selection Practices and Maize Variety Characteristics in a Traditionally-Based Mexican Community. CIMMYT Economics Working Paper No. 98-04. Mexico, D. F.: CIMMYT.

Loy, D. D., & Lundy, E. L. (2019). Nutritional properties and feeding value of corn and its coproducts. In: Serna-Saldivar, S. O., ed. *Corn* (Third Edition). Oxford: AACC International Press, 633–659.

Luna, V. S., Figueroa, M. J., Baltazar, M. B., Gomez, L. R., Townsend, R., & Schoper, J. B. (2001). Maize pollen longevity and distance isolation requirements for effective pollen control. *Crop Science*, 41, 1551–1557.

Matsuoka, Y., Vigouroux, Y., Goodman, M. M., Sanchez, G. J., Buckler, E., & Doebley, J. (2002). A single domestication for maize shown by multilocus microsatellite genotyping. *Proceedings of the National Academy of Science U S A*, 99(9), 6080–6084. doi: 10.1073/pnas.052125199

Mangelsdokf, P. C., & Reeves, R. G. (1939). The origin of Indian corn and its relatives, 315.

Mangelsdorf, P. C. (1974) *Corn: Its origin evolution and improvement*. The Belknap Press of Harvard University, Cambridge.

Mann, K. K. (2014). Health benefits of Maize phytochemicals. *The Journal of Applied Social Science* , *1*(2&3), 107–114.

Mao, H., Wang, H., Liu, S., Li, Z., Yang, X., & Yan, J. et al. (2015). A transposable element in a NAC gene is associated with drought tolerance in maize seedlings. *Nature Communications*, 6(1), 13. doi: 10.1038/ncomms9326

Marks, B. B., Megías, M., Ollero, F. J., Nogueira, M. A., Araujo, R. S., & Hungria, M. (2015). Maize growth promotion by inoculation with Azospirillum brasilense and metabolites of Rhizobium tropici enriched on lipo-chitooligosaccharides (LCOs). *AMB Express*, 5, 1–11. doi: 10.1186/s13568-015-0154-z

Mann, K. K. (2014). Health benefits of Maize phytochemicals. *The Journal of Applied Social Science*, 1(2&3), 107–114.

Miao, Z., Han, Z., Zhang, T., Chen, S., & Ma, C. (2017). A systems approach to a spatio-temporal understanding of the drought stress response in maize. *Scientific Reports*, 7, 1–14. doi: 10.1038/s41598-017-06929-y

Monjardino, P., Smith, A. G., & Jones, R. J. (2005). Heat stress effects on protein accumulation of maize endosperm. *Crop Science*, 45, 1203–1210.

Monjardino, P., Smith, A. G., & Jones, R. J. (2006). Zein transcription and endoreduplication in maize endosperm are differentially affected by heat stress. *Crop Science*, 46, 25.

Montañez, A., Blanco, A. R., Barlocco, C., Beracochea, M., & Sicardi, M. (2012). Characterization of cultivable putative endophytic plant growth promoting bacteria associated with maize cultivars (Zea mays L.) and their inoculation effects in vitro. *Applied Soil Ecology*, 58, 21–28. doi: 10.1016/j.apsoil.2012.02.009

Milind, P., & Isha, D. (2013). Zea maize: A modern craze. *International Research Journal of Pharmacy*, 4, 39–43.

Mottet, A., de Haan, C., Falcucci, A., Tempio, G., Opio, C., & Gerber, P. (2017). Livestock: On our plates or eating at our table? A new analysis of the feed/food debate. *Global Food Security*, 14, 1–8.

Mukherjee, B. K. (1989). Cross-pollinated Crops: Maize. *Plant Breeding: Theory and Practice*, 199.

Mukherjee, B. K. 1989. Maize. In: Chopra, V. L., ed. *Plant breeding*. New Delhi: Oxford & IBH, 199–212.

Mukherjee, B. K., & Dhawan, N. L. (1970). Genetic studies in varieties of maize. I. Genetic analysis of yield and other quantitative characters. *Japanese Journal of Breeding*, 20, 47–56.

National Research Council. (1930). Quality-Protein Maize: Report of an Ad Hoc Panel of the Advisory Committee on Technology Innovation Board on Science and Technology for International Development National Research Council, in Cooperation With the Board on Agriculture National Research Co.

Nielsen, K. M., Gebhard, F., Smalla, K., Bones, A. M., & van Elsas, J. D. (1997). Evaluation of possible horizontal gene transfer from transgenic plants to the soil bacterium *Acinobactor calcoaceticus* BD413. *Theoretical and Applied Genetics*, 95, 815–821.

Nielsen, K. M., Bones, A. M., Smalla, K., & van Elsas, J. D. (1998). Horizontal gene transfer from transgenic plants to terrestrial bacteria - a rare event. *FEMS Microbiology Reviews*, 22, 79–103.

Nuss, E. T., & Tanumihardjo, S. A. (2010) Maize: A paramount staple crop in the context of global nutrition. 9(4), 417–436. doi: 10.1111/j.1541-4337.2010.00117.x

Nur, H. B., Michaela, L., Made, S., Josh, V. V., Peter, A., & Amy, I. et al. (2020). Meeting the food security challenge for nine billion people in 2050: What impact on forests? *Global Environmental Change*, 65, 102195. doi: 10.1016/j.gloenvcha.2020.102056

Oaks, A., & Beevers, H. (1964). The glyoxylate cycle in maize scutellum. *Plant Physiology*, 39, 431–434.

Odell, S. G., Hudson, A. I., Praud, S., Dubreuil, P., Tixier, M. H., & Ross-Ibarra, J. et al. (2022). Modeling allelic diversity of multi-parent mapping populations affects detection of quantitative trait loci. *G3 (Bethesda)*, 12(3), 1–14. doi: 10.1093/g3journal/jkac011

OECD. (2002). Consensus Document on compositional considerations for new varieties of maize (Zea mays): key food and feed nutrients, anti-nutrients and secondary plant metabolites. *Series on the Safety of Novel Food and Feeds (ENV/JM/MONO (2002) 25)*, 6, 1–42.

Okoruwa, A. E., & Kling, J. G. (1996). Nutrition and Quality of Maize. International Institute of Tropical Agriculture, Research Guide 33, available online at http://www.iita.org/cms/details/trn_mat/irg33/irg33.html

Pasini, G., Simonato, B., Curioni, A., Vincenzi, S., Cristaudo, A., Santucci, B., Peruffo, A. D., & Giannattasio, M. (2002). IgE-mediated allergy to corn: A 50 kDa protein, belonging to the Reduced Soluble Proteins, is a major allergen. *Allergy*, 57, 98–106.

Pastorello, E. A., Farioloi, L., Pravettoni, V., Ispano, M., Scibola, E., Trambaioli, C., Giuffrida, M. G., Ansaloni, R., Godovac-Zimmermann, J., Conti, A., Fortunato, D., & Ortolani, C. (2000). The maize major allergen, which is responsible for food-induced allergic reactions, is a lipid transfer protein. *Journal of Allergy and Clinical Immunology*, 106, 744–751.

Paulsen, M. R., Singh, M., & Singh, V. (2019). Measurement and maintenance of corn quality. In: Serna-Saldivar, S. O., ed. *Corn* (Third Edition). Oxford: AACC International Press, 165–211.

Pavelchak, N. (1999). Silo gas exposure in New York State following the dry growing season of 1995. *Applied Occupational and Environmental Hygiene*, 14, 34–38.

Peña-Rosas, J. P., Garcia-Casal, M. N., Pachón, H., McLean, M. S., & Arabi, M. (2014). Technical considerations for maize flour and corn meal fortification in public health: Consultation rationale and summary. *Annals of the New York Academy of Sciences*, 1312, 1–7. doi: 10.1111/nyas.12434

Pérez-Montaño, F., Jiménez-Guerrero, I., Sánchez-Matamoros, R. C., López-Baena, F. J., Ollero, F. J., & Rodríguez-Carvajal, M. A. et al. (2013). Rice and bean AHL-mimic quorum-sensing signals specifically interfere with the capacity to form biofilms by plant associated bacteria. *Research in Microbiology*, 164, 749–760. doi: 10.1016/j.resmic.2013.04.001

Poehlman, J. M. (1959). Breeding field crops. pp. 147–163.

Prasanna, B. M., Huesing, J. E., Eddy, R., & Peschke, V. M. (Eds.). (2018). Fall Armyworm in Africa: A Guide for Integrated Pest Management. CIMMYT, Mexico, CDMX. https://repository.cimmyt.org/handle/10883/19204

Prasanna, B. M., Palacios-Rojas, N., Hossain, F., Muthusamy, V., Menkir, A., Dhliwayo, T., & Fan, X. (2020). Molecular breeding for nutritionally enriched maize: Status and prospects. *Frontiers in Genetics*, 10, 1392.

Prasanna, B. M., Cairns, J. E., Zaidi, P. H., Beyene, Y., Makumbi, D., Gowda, M., Magorokosho, C., Zaman-Allah, M., Olsen, M., Das, A., Worku, M., Gethi, J., Vivek, B. S., Nair, S. K., Rashid, Z., Vinayan, M. T., Issa, A. B., San Vicente, F., Dhliwayo, T., & Zhang, X. (2021). Beat the stress: breeding for climate resilience in maize for the tropical rainfed environments. *Theoretical and Applied Genetics*, 134(6), 1729–1752. doi: 10.1007/s00122-021-03773-7

Prasanna, B. M., Pixley, K., Warburton, M. L., & Xie, C. X. (2009). Molecular marker-assisted breeding options for maize improvement in Asia. *Molecular Breeding*, 26(2), 339–356.

Purseglove, J. W. (1972). *Tropical crops: Monocotyledons 1*, London, Longman Group Limited.

Qu, J., Zhong, Y., Ding, L., Liu, X., Xu, S., Guo, D., & Xue, J. (2022). Biosynthesis, structure and functionality of starch granules in maize inbred lines with different kernel dehydration rate. *Food Chemistry*, 368, 130796.

Radu, A., Urechean, V., Naidin, C., & Motorga, V. (1997). The theoretical significance of a mutant in the phylogeny of the species *Zea mays* L. *Maize Newsletter*, 71, 77–78.

Raynor, G. S., Ogden, E. C., & Hays, J. V. (1972). Dispersion and deposition of corn pollen from experimental sources. *Agronomy Journal*, 64, 420–427.

Roberts, M. J. (2006). The value of plant disease early-warning systems: a case study of USDA's soybean rust coordinated framework (No. 18) (USDA: Economic Research Service).

Rhoades, M. M. (1931). Cytoplasmic inheritance of male sterility in Zea mays. *Science*, 73(1891), 340–341.

Sammon, A. M., & Iputo, J. E. (2006). Maize meal predisposes to endemic squamous cancer of the oesophagus in Africa: Breakdown of esterified linoleic acid to the free form in stored meal leads to increased intragastric PGE2 production and a low-acid reflux. *Medical Hypotheses*, 6, 1431–1436.

Schlüter, K., Fütterer, J., & Potrykus, I. (1995). "Horizontal" gene transfer from a transgenic potato line to a bacterial pathogen (*Erwinia chrysanthemi*) occurs—if at all—at an extremely low frequency. *Bio/technology*, 13(10), 1094–1098.

Senaphan, K., Kukongviriyapan, U., Sangartit, W., Pakdeechote, P., Pannangpetch, P., & Prachaney, P. et al. (2015). Ferulic acid alleviates changes in a rat model of metabolic syndrome induced by high-carbohydrate, high-fat diet. *Nutrients*, 7(8), 6446–6464. doi: 10.3390/nu7085283

Serna-Saldivar, S. O., & Perez Carrillo, E. (2019). Food uses of whole corn and dry-milled fractions. In: Serna-Saldivar, S. O., ed. *Corn* (Third Edition). Oxford: AACC International Press, 435–467.

Sharma, R. C., Payak, M. M., & Sachan, J. K. S. 1995. Disease reaction of wild relatives of maize to *Erwinia* stalk rot and *Helminthosporium* leaf blight. 69: 111–112.

Shiferaw, B., Prasanna, B. M., Hellin, J., & Bänziger, M. (2011). Crops that feed the world 6. Past successes and future challenges to the role played by maize in global food security. *Food Security*, 3(3), 307. doi: 10.1007/s12571-011-0140-5

Singh, B. (1977). *Races of maize in India*, New Delhi, ICAR.

Sitara, U., & Akhter, S. (2007). Efficacy of fungicides, sodium hypochloritc and neem seed powder to control seed borne pathogens of maize. *Pakistan Journal of Botany*, 39(1), 285.

Stein, A. J., & Rodriguez-Cerezo, E. (2009). JRC Scientific and Technical Reports.

Staiger, C. J., Goodbody, K. C., Hussey, P. J., Valenta, R., Drøbak, B. K., & Lloyd, C.W. (1993). The profilin multigene family of maize: differential expression of three isoforms. *Plant Journal*, 4(4), 631–641. doi: 10.1046/j.1365-313x.1993.04040631.x

Syvanen, M. (1999). In search of horizontal gene transfer. *Nature Biotechnology*, 17, 833.

Talukder, Z. A., Muthusamy, V., Chhabra, R., Bhatt, V., Reddappa, S. B., Mishra, S. J., ... & Hossain, F. (2022). Enrichment of amylopectin in sub-tropically adapted maize hybrids through genomics-assisted introgression of waxy1 gene encoding granule-bound starch synthase (GBSS). *Journal of Cereal Science*, 105, 103443.

Tanumihardjo, S. A., McCulley, L., Roh, R., Lopez-Ridaura, S., Palacios-Rojas, N., & Gunaratna, N. S. (2020). Maize agro-food systems to ensure food and nutrition security in reference to the Sustainable Development Goals. *Global Food Security*, 25, 100327.

Tesfaye, K., Gbegbelegbe, S., Cairns, J. E., Shiferaw, B., Prasanna, B. M., Sonder, K., ... & Robertson, R. (2015). Maize systems under climate change in sub-Saharan Africa: Potential impacts on production and food security. *International Journal of Climate Change Strategies and Management*, 7(3), 247–271.

Tetlow, I. J. (2011). Starch biosynthesis in developing seeds. *Seed Science Research*, 21(1), 5–32.

Tigar, B. J., Osborne, P. E., Key, G. E., Flores-S, M. E., & Vazquez-A, M. (1994). Insect pests associated with rural maize stores in Mexico with particular reference to Prostephanus truncatus (Coleoptera: Bostrichidae). *Journal of Stored Products Research*, 30, 267–281. doi: 10.1016/S0022-474X(94)90319-0

Tittonell, P., Gérard, B., & Erenstein, O. (2015). Tradeoffs around crop residue biomass in smallholder crop-livestock systems – What's next? *Agricultural Systems*, 134, 119–128.

Trinidad-Calderon, P. A., Acosta-Cruz, E., Rivero-Masante, M. N., Diaz- Gomez, J. L., Garcia-Lara, S., & Lopez-Castillo, L. M. (2021). Maize bioactive peptides: From structure to human health. *Journal of Cereal Science*, 100, 103232. doi: 10.1016/j.jcs.2021.103232

Tripathi, K. K., Warrier, R., Govilla, O. P., & Ahuja, V. 2011. Biology of *Zea mays* (Maize). Department of Biotechnology, Ministry of Science and Technology and Ministry of environment and forests, Govt. of India, 2011.

Tunwar, N. S., & Singh, S. V. 1988. Indian Minimum Seed Certification Standards. The Central Seed Certification Board, Department of Agriculture and Cooperation, Ministry of Agriculture, Government of India.

Tubiello, F. N., Mayo, R., & Salvatore, M. (2009). Agriculture and greenhouse gases: FAO's approach to addressing the unique challenges faced by agricultural statisticians. *Rome, Italy, Viale delle Terme di Caracalla.*

Tziotis, A., Seetharaman, K., Klucinec, J. D., Keeling, P., & White, P. J. (2005). Functional properties of starch from normal and mutant corn genotypes. *Carbohydrate Polymers*, 61(2), 238–247.

Ufaz, S., & Galili, G. (2008). Improving the content of essential amino acids in crop plants: goals and opportunities. *Plant Physiology*, 147, 954–961.

UN-DESA. (2019). *World population prospects 2019*, New York, UN Department for Economic and Social Affairs.

Valbuena, D., Erenstein, O., Homann-Kee Tui, S., Abdoulaye, T., Claessens, L., Duncan, A. J., Gerard, B., Rufino, M. C., Teufel, N., van Rooyen, A., & van Wijk, M. T. (2012). Conservation Agriculture in mixed crop-livestock systems: Scoping crop residue tradeoffs in sub-Saharan Africa and South Asia. *Field Crops Research*, 132, 175–184.

3.2 CENTER OF ORIGIN AND EARLY DOMESTICATION

Maize (*Zea mays* ssp. *mays*) was domesticated from the wild grass *Zea mays* ssp. *parviglumis*, commonly known as teosinte, in the Balsas River Valley of southwestern Mexico approximately 9,000 years ago (Matsuoka et al., 2002; Piperno et al., 2009). This region is considered the primary center of origin due to a combination of archaeological, morphological, and molecular genetic evidence that traces the lineage of modern maize back to this geographic and genetic source. Initial domestication involved the transformation of teosinte, which bears few cobs with small, hard kernels, into the multi-rowed, soft-kernelled maize phenotype recognizable today. This domestication process likely occurred over centuries through unconscious and conscious selection by early Mesoamerican farmers for traits such as non-shattering ears, larger kernel size, softer pericarps, and more robust cob structure (Doebley et al., 2006). Genetic studies have demonstrated that relatively few genomic regions were involved in the initial domestication process, but these had large phenotypic effects, particularly genes such as *tb1* (teosinte branched1), which controls apical dominance and plant architecture (Studer et al., 2011).

Archaeobotanical findings, including starch grains and phytoliths from ancient grinding stones, have confirmed the presence of domesticated maize in the Balsas region by at least 8,700 years before present (Piperno et al., 2009). Moreover, subsequent diffusion and diversification of maize were facilitated by early human migrations, leading to the emergence of diverse landraces adapted to varying ecological zones in Central and South America (Van Heerwaarden et al., 2011; Kistler et al., 2018). This domestication event not only represents one of the most successful plant domestication stories in human history but also laid the foundation for the later adaptation of maize to an extraordinarily broad range of environments—including the temperate zones far from its tropical origins.

3.3 PATHWAYS OF DISPERSAL

Following its domestication in the Balsas River Valley of southwestern Mexico approximately 9,000 years ago, maize began to spread both northward and southward across the Americas, eventually becoming one of the most widely cultivated crops in pre-Columbian societies (Matsuoka et al., 2002; Piperno et al., 2009). This dispersal was facilitated primarily by human migration, trade networks, and cultural exchange, allowing maize to adapt to a wide range of ecological zones from the high Andes to the eastern woodlands of North America.

3.3.1 Southward Expansion into Central and South America

Maize reached Panama by around 7,000 years ago, as evidenced by phytolith and starch grain remains in archaeological contexts (Piperno & Flannery, 2001). Its subsequent movement into South America is believed to have occurred via both Pacific coastal and Andean highland routes, where maize underwent further diversification under different environmental and cultural selection pressures (Kistler et al., 2018). Genomic studies suggest that South American maize underwent significant adaptation and even introgression with wild relatives, which contributed to its diversification into locally adapted landraces (Van Heerwaarden et al., 2011; Grobman et al., 2012).

3.3.2 Northward Movement into North America

Maize dispersed northward into what is now the southwestern United States by approximately 4,100 years ago, where it initially remained a minor component of the diet (Merrill et al., 2009). The full agricultural reliance on maize in the U.S. Southwest occurred around 2,000–3,000 years ago, as suggested by archaeological and isotopic studies (Roth et al., 2018). The crop then spread eastward into the Mississippi Valley and the eastern woodlands, where regional varieties such as Northern Flint maize emerged in response to cooler and shorter growing seasons (Radin, 2006; da Fonseca et al., 2015).

3.3.3 Transoceanic Dispersal

After the Columbian Exchange began in the late 15th century, maize was rapidly disseminated across Europe, Africa, and Asia by European explorers, traders, and colonizers (Mir et al., 2013). Maize was introduced to Spain by the early 1500s and quickly spread throughout the Mediterranean region, reaching sub-Saharan Africa, India, and China within the next few centuries (Morris et al., 2013). In each new region, farmers selected for traits suited to local conditions, which led to the formation of a broad array of temperate and tropical landraces worldwide.

3.3.4 Genetic Evidence Supporting Dispersal Pathways

Population genomic analyses have revealed that maize landraces retain genetic signatures of their dispersal routes. North American and South American landraces show distinct ancestry patterns, with signs of introgression from wild relatives and evidence of selection for traits such as flowering time and drought tolerance (Hufford et al., 2012). These findings underscore the dynamic interplay between human-mediated movement and environmental adaptation in shaping the genetic landscape of maize.

3.4 CHALLENGES IN ADAPTING TO TEMPERATE CLIMATES

The migration of maize from its tropical center of origin into temperate regions posed significant adaptive challenges. Unlike its native environment—characterized by relatively constant photoperiods, warm temperatures, and long growing seasons—temperate zones present markedly different climatic regimes, including shortened growing seasons, cooler temperatures, unpredictable frosts, and long-day photoperiods (Hung et al., 2012; Swarts et al., 2017).

3.4.1 Photoperiod Sensitivity and Flowering Time

One of the foremost barriers to successful temperate adaptation was photoperiod sensitivity. Tropical maize is typically short-day flowering, initiating reproductive development in response to decreasing day length. In contrast, temperate environments, especially at higher latitudes, experience longer days during the growing season, which initially delayed flowering in tropical genotypes (Coles et al., 2010). Natural and artificial selection acted strongly on flowering time regulatory loci, particularly ZCN8, Vgt1, and ZmCCT, leading to the evolution of genotypes with reduced photoperiod sensitivity and earlier flowering—traits essential for completing the life cycle within the limited growing season of temperate climates (Hung et al., 2012; Guo et al., 2018).

3.4.2 Cold Sensitivity and Germination Constraints

Maize is inherently sensitive to cold stress, especially during seed germination and early seedling establishment. Tropical genotypes often exhibit poor emergence and reduced vigor under low soil temperatures common in temperate spring conditions. This challenge has constrained the expansion of maize into areas with cooler climates and necessitated the development of cold-tolerant genotypes through selection and breeding (Revilla et al., 2016). Low temperature impairs cell division, membrane fluidity, and enzyme activity, thereby affecting early root and shoot development (Fracheboud et al., 2004).

3.4.3 Frost Susceptibility and Short Season Constraints

Unlike cold-tolerant cereals such as wheat or barley, maize lacks inherent frost resistance. Both vegetative and reproductive tissues are highly vulnerable to freezing temperatures, limiting the length of the growing window. In temperate climates, especially those with a risk of early autumn frosts, maize must complete its life cycle rapidly. This pressure has favored the selection of early maturing varieties with rapid grain filling and synchronized development (Tenaillon & Charcosset, 2011).

3.4.4 Nutrient Use and Soil Adaptation

Maize's adaptation to temperate soils also posed agronomic challenges. In many temperate zones, soils are more acidic or nutrient-limited compared to the fertile volcanic or alluvial soils in the tropics. Cold temperatures further reduce nutrient mineralization and uptake efficiency. Over time, temperate maize genotypes have been selected for enhanced root development, efficient nitrogen uptake, and early phosphorus acquisition, traits crucial for performance in cool soils (Zhao et al., 2006).

3.4.5 Pathogen and Pest Pressures

As maize moved into new agroecological zones, it encountered novel pathogens and pests, such as northern corn leaf blight (*Exserohilum turcicum*) and the European corn borer (*Ostrinia nubilalis*), which were absent or less prevalent in tropical regions. The need to maintain yield under biotic stress led to the incorporation of host resistance genes and agronomic strategies for integrated pest management (Meihls et al., 2012).

3.5 GENETIC AND PHYSIOLOGICAL ADAPTATIONS

Maize's successful establishment in temperate climates is a product of complex genetic and physiological modifications driven by both natural and artificial selection (Table 3.1)

These adaptations have enabled maize to cope with low temperatures, variable photoperiods, short growing seasons, and novel stressors. The integration of favorable alleles through recurrent selection, introgression, and mutation has resulted in phenotypes capable of performing under temperate conditions without compromising yield potential (Romay et al., 2013; Li et al., 2013).

3.5.1 Genetic Modifications in Flowering Time Regulation

The shift from short-day photoperiodism to day-neutral or long-day tolerance was a cornerstone of temperate adaptation. Several quantitative trait loci (QTLs) and major genes regulating flowering time have undergone strong selection. The ZmCCT gene, which inhibits flowering under long days, shows reduced expression in temperate lines due to transposon insertions in its promoter region—thereby accelerating flowering under long photoperiods (Hung et al., 2012). Similarly, allelic variants of Vgt1, a cis-regulatory region upstream of the ZmRap2.7 gene, have been associated with early flowering in European and North American germplasm (Salvi et al., 2007).

3.5.2 Cold Tolerance and Seedling Vigor

Physiologically, one of the key challenges in temperate zones is the ability of maize to germinate and establish under suboptimal soil temperatures. Genotypes adapted to temperate regions often exhibit improved membrane stability, osmolyte accumulation, and antioxidant enzyme activity under cold conditions (Goering et al., 2021).

TABLE 3.1
Key Genetic Loci Involved in the Adaptation of Maize to Temperate Climates

Gene/Locus	Chromosome	Function	Role in Adaptation	Reference
ZmCCT	Chr10	Photoperiod regulation	Reduction in daylength sensitivity	Yang et al., 2013
Vgt1	Chr8	Flowering time control	Promotes early flowering under long-day conditions	Salvi et al., 2007
ZCN8	Chr8	Florigen signal gene	Regulates transition to flowering; key for adaptation to temperate photoperiods	Meng et al., 2011
id1	Chr1	Developmental timing	Controls flowering initiation via sugar signaling	Colasanti et al., 1998
Dwarf8	Chr1	Growth and flowering	Affects plant height and flowering time under temperate conditions	Thornsberry et al., 2001

These traits enhance root and shoot growth under early season chilling. Genetic studies have identified several loci and transcription factors involved in cold response pathways, including CBF/DREB-like genes, which modulate cold-induced gene expression in maize (Leng et al., 2014).

3.5.3 Leaf and Canopy Physiology

Adapted temperate maize also displays physiological traits that enhance photosynthetic efficiency under fluctuating light and temperature regimes. These include modifications in stomatal conductance, chlorophyll content, and rubisco activation, which help sustain carbon fixation during cooler mornings and shorter growing periods. Some temperate-adapted lines also exhibit altered canopy architecture, such as more upright leaves, which optimize light interception in high-density planting common in cooler climates (Hammer et al., 2009).

3.5.4 Hormonal and Metabolic Adjustments

Hormonal responses play a critical role in stress mitigation and developmental synchronization. Temperate maize lines show different profiles of abscisic acid (ABA), gibberellins, and cytokinins, which regulate stress response and flowering (Qin et al., 2011). Metabolomic studies have shown shifts in sugar metabolism, amino acid pools, and lipid composition, which support cell stability and energy balance under cold and nutrient-limited conditions (Riva-Roveda et al., 2016).

3.5.5 Epigenetic and Transcriptomic Plasticity

Recent advances highlight that temperate adaptation also involves epigenetic regulation and transcriptional plasticity. Differential DNA methylation and small RNA expression patterns have been observed between tropical and temperate maize, influencing traits like flowering and stress response (Eichten et al., 2013). Transcriptomic studies further reveal that temperate genotypes possess enhanced transcriptional responsiveness to low temperature, supporting rapid acclimation (Avila et al., 2018).

3.6 ROLE OF HUMAN SELECTION AND BREEDING

The transition of maize from a tropical to a temperate crop is inseparable from the role of human agency. While natural selection acted on maize populations exposed to new environmental pressures, it was human selection and breeding—both traditional and modern—that accelerated and stabilized adaptive traits. Over thousands of years, farmers and breeders have exerted selection pressure on flowering time, maturity, cold tolerance, and plant architecture, and grain characteristics, progressively refining maize to perform in temperate agroecosystems.

3.6.1 Early Farmer Selection and Local Adaptation

Pre-Columbian farmers were instrumental in shaping maize's local adaptation across diverse environments in Mesoamerica and, later, North and South America.

As maize spread geographically, early cultivators selected plants that flowered earlier, matured faster, or yielded reliably under local conditions (Van Heerwaarden et al., 2011). This decentralized, farmer-driven selection process created a wide array of landraces—genetically diverse populations adapted to specific altitudes, latitudes, and climates (Pressoir & Berthaud, 2004). These landraces laid the foundation for further breeding and provided essential alleles for adaptation to cold and photoperiodic variation.

3.6.2 Modern Breeding for Temperate Adaptation

The advent of scientific plant breeding in the 20th century catalyzed rapid genetic improvements in maize. In temperate zones like the U.S. Corn Belt and Europe, breeders selected for shorter time to flowering, higher yield under cooler conditions, and resistance to abiotic and biotic stresses (Duvick, 2005). Recurrent selection, hybrid breeding, and pedigree selection enabled the accumulation of favorable alleles while purging deleterious ones.

The development of double-cross and single-cross hybrids in temperate regions dramatically improved performance stability and heterosis (Hallauer et al., 2010). Notably, modern breeding also capitalized on exotic germplasm—reintroducing alleles from tropical landraces or wild relatives to improve cold tolerance and nutrient use efficiency (Troyer, 2004).

3.6.3 Molecular Breeding and Genomics-Assisted Selection

In recent decades, breeding strategies have become increasingly genotype-informed. Genomic tools such as marker-assisted selection (MAS), genomic selection (GS), and QTL mapping have accelerated the identification of loci underlying traits crucial for temperate adaptation—e.g., flowering time (Vgt1, ZmCCT), cold tolerance, and root architecture (Xu et al., 2020).

With the sequencing of the maize genome (Schnable et al., 2009), breeders gained access to a wealth of functional and structural variation, enabling precise introgression and selection. GS models, particularly in temperate breeding programs, have improved accuracy and reduced breeding cycle time, allowing faster release of climate-resilient varieties (Crossa et al., 2017).

3.6.4 Participatory Breeding and Local Innovations

In some temperate regions, especially in developing countries or marginal environments, participatory breeding has been instrumental in adapting maize to localized temperate niches. These programs integrate farmer knowledge, local preferences, and adaptive breeding, often focusing on traits like earliness, stress tolerance, and culinary quality (Ceccarelli & Grando, 2007).

This approach sustains genetic diversity and ensures that breeding products are socially relevant and environmentally suited, bridging the gap between global science and local practice.

3.7 IMPLICATIONS OF CLIMATE CHANGE AND FUTURE BREEDING

The adaptation of maize (*Zea mays* L.) to temperate climates is a historical testament to its remarkable plasticity and the synergistic role of environmental selection and human intervention. However, as global climate change accelerates, maize faces new ecological and agronomic constraints that demand both innovation and foresight. Rising atmospheric temperatures, erratic precipitation, and increasing frequency of abiotic and biotic stresses are predicted to significantly impact maize production across both tropical and temperate zones. This section evaluates the potential implications of climate change on maize cultivation in temperate regions, drawing from physiological, ecological, and breeding perspectives, and outlines strategic predictions and pathways for future resilience.

3.7.1 Warming Temperatures and Shifting Phenology

One of the most direct effects of climate change is the elevation in mean and extreme temperatures. In temperate maize-growing regions, even modest increases in temperature can have disproportionately large impacts on phenological development. Maize, being a thermophilic crop, requires warm conditions for optimal growth; however, high temperatures during reproductive stages, particularly anthesis and silking, can cause asynchronous flowering, pollen sterility, and kernel abortion, ultimately reducing yield (Lobell et al., 2011; Hatfield & Prueger, 2015).

Moreover, higher temperatures can lead to shortened growing periods, affecting biomass accumulation and grain filling. Studies suggest that with every 1°C rise in global temperature, maize yields may decline by 7–10%, with larger reductions in already marginal or rainfed environments (Zhao et al., 2017). For temperate climates, the effect is compounded by the narrow thermal tolerance of many elite maize cultivars, originally bred for stable environmental conditions.

3.7.2 Drought and Hydrological Variability

The increased unpredictability of precipitation—ranging from prolonged droughts to intense, short-duration rainfall events—poses another major challenge. In temperate maize systems, especially those dependent on rainfed agriculture, water availability during vegetative growth and flowering is crucial. Drought during silking leads to poor pollination, while water stress during grain filling reduces starch accumulation, severely impacting yield (Daryanto et al., 2016).

Climate projections indicate reduced snowpack in high-latitude regions and earlier onset of snowmelt, which could result in early season water deficits. Maize varieties with traits like deep and robust root systems, efficient stomatal control, and osmotic adjustment will be essential in adapting to these shifts.

3.7.3 Biotic Stress Expansion: Pests and Pathogens

Changing climatic conditions are altering the geographic distribution of insect pests and diseases, many of which are migrating northward into temperate zones. The warming trend facilitates the overwintering survival of pest populations, allowing for earlier and more aggressive infestations. Notably, corn rootworm (Diabrotica spp.), European corn borer (*O. nubilalis*), and southern rust (*Puccinia polysora*) are expected to expand into new areas with climate warming (Oerke, 2006). Additionally, elevated CO_2 concentrations may affect host-pathogen dynamics, potentially enhancing fungal virulence or weakening plant defense mechanisms. Consequently, future breeding efforts must focus on durable resistance genes, multi-pathogen resistance, and integrated pest management systems tailored for dynamic environments.

3.7.4 Breeding Strategies for Climate Resilience

The historical success of maize adaptation to diverse climates provides a valuable blueprint for current and future breeding programs. Modern climate-resilient maize must combine tolerance to abiotic stresses (heat, drought, salinity) with resistance to biotic threats and yield stability across environments. Recent advances include:

- **GS and prediction** for performance under multiple stress conditions (Crossa et al., 2017).
- **CRISPR-Cas9 genome editing** to target specific alleles controlling flowering time, stomatal function, and heat stress (Varshney et al., 2021).
- Identification and deployment of **QTLs** for cold tolerance, such as *ZmCCT* and *Vgt1*, which can accelerate adaptation to new temperate areas (Xu et al., 2020).
- Use of **wild relatives and landraces** to introduce rare alleles for stress adaptation without compromising agronomic performance.

Furthermore, speed breeding and high-throughput phenotyping platforms are enabling faster selection cycles, allowing breeders to respond more rapidly to climate-induced challenges

3.7.5 Geographic Shifts and Expansion Opportunities

While climate change presents significant challenges, it may also open up new areas for maize cultivation as warming temperatures render previously inhospitable regions viable. High-latitude regions in northern Europe, southern Canada, and eastern Russia are projected to see longer growing seasons and increased growing degree days, making maize expansion feasible (Monfreda et al., 2008). However, successful expansion into new zones will require:

- Region-specific breeding programs to develop early maturing and cold-tolerant cultivars.
- Adaptive agronomy, including modified planting dates, fertilizer regimes, and crop rotations.
- Supportive policy and infrastructure, such as cold-resistant seed systems and farmer training in new regions.

3.7.6 Socioeconomic and Food Security Dimensions

The impact of climate change on maize goes beyond agronomic performance—it also has far-reaching implications for food security, especially in regions that rely on maize as a staple. Yield instability and crop failure in temperate regions can disrupt global grain markets and exacerbate food insecurity in import-dependent countries. Equitable access to climate-smart technologies, inclusive policy frameworks, and participatory breeding involving local farmers will be crucial to ensure that the benefits of scientific advancements reach smallholders, who are often most vulnerable to climate shocks (Ceccarelli & Grando, 2007).

3.8 CONCLUSION

The journey of maize (*Zea mays* L.) from its tropical origin in southern Mexico to its widespread cultivation in temperate agroecological zones is a profound example of plant adaptation, human ingenuity, and evolutionary resilience. From its initial domestication nearly 9,000 years ago to its global diffusion via diverse geographic, cultural, and ecological pathways, maize has undergone extensive transformations—morphological, physiological, genetic, and agronomic. The successful adaptation of maize to temperate climates involved a complex interplay of natural selection and deliberate human breeding. Key transitions included modifications in photoperiod sensitivity, flowering time regulation, and cold tolerance. These adaptations were supported by selection at major loci such as *ZCN8*, *Vgt1*, and *ZmCCT*, among others, alongside cumulative changes in quantitative traits influenced by polygenic selection. Early agricultural societies and, later, modern breeding programs played pivotal roles in shaping maize into a crop suitable for short growing seasons, cooler temperatures, and new pest and disease pressures.

The evolutionary narrative of maize highlights its immense genetic plasticity, which has allowed it to thrive in an astonishing range of environments. However, this resilience is now being tested by climate change, with rising temperatures, water scarcity, and emerging biotic stresses threatening maize productivity, especially in high-latitude regions. At the same time, shifting climatic zones present new opportunities for maize expansion into formerly marginal areas. To ensure the continued success of maize in temperate zones—and globally—future efforts must prioritize climate-smart breeding strategies, incorporating genomic tools, gene editing technologies, and transdisciplinary approaches. These must be complemented by adaptive agronomic practices, supportive policy frameworks, and inclusive innovation systems that serve both commercial agriculture and smallholder farming communities. Ultimately, the history of maize offers a valuable blueprint for navigating future uncertainties. Just as maize once traversed ecological boundaries through evolutionary adaptation and human vision, its continued success in a changing world will depend on our collective capacity to innovate, conserve genetic diversity, and steward this essential crop toward sustainable resilience.

REFERENCES

Avila, L. M., Obeidat, W., & Aleem, A. A. (2018). Cold stress response mechanisms in maize: From gene expression to breeding. *Journal of Crop Improvement*, 32(5), 571–589.

Ceccarelli, S., & Grando, S. (2007). Decentralized-participatory plant breeding: an example of demand driven research. *Euphytica*, 155(3), 349–360.

Coles, N. D., McMullen, M. D., Balint-Kurti, P. J., Holland, J. B., & Pratt, R. C. (2010). Genetic control of photoperiod sensitivity in maize revealed by joint multiple population analysis. *Genetics*, 184(3), 799–812. https://doi.org/10.1534/genetics.109.110536

Colasanti, J., Yuan, Z., & Sundaresan, V. (1998). The indeterminate gene encodes a zinc finger protein and regulates a leaf-generated signal required for the transition to flowering in maize. *Cell*, 93(4), 593–603.

Crossa, J., Pérez-Rodríguez, P., Cuevas, J., Montesinos-López, O., Jarquín, D., de los Campos, G., & Hickey, J. M. (2017). Genomic selection in plant breeding: Methods, models, and perspectives. *Trends in Plant Science*, 22(11), 961–975. https://doi.org/10.1016/j.tplants.2017.08.011

da Fonseca, R. R., Smith, B. D., Wales, N., Cappellini, E., Skoglund, P., Fumagalli, M., & Gilbert, M. T. P. (2015). The origin and evolution of maize in the Southwestern United States. *Nature Plants*, 1(9), 1–5. https://doi.org/10.1038/nplants.2015.110

Daryanto, S., Wang, L., & Jacinthe, P. A. (2016). Global synthesis of drought effects on maize and wheat production. *PLOS ONE*, 11(5), e0156362. https://doi.org/10.1371/journal.pone.0156362

Doebley, J. F., Gaut, B. S., & Smith, B. D. (2006). The molecular genetics of crop domestication. *Cell*, 127(7), 1309–1321. https://doi.org/10.1016/j.cell.2006.12.006

Duvick, D. N. (2005). Genetic progress in yield of United States maize (Zea mays L.). *Maydica*, 50(3/4), 193.

Eichten, S. R., Schmitz, R. J., & Springer, N. M. (2013). Epigenetics: Beyond chromatin modifications and complex genetic regulation. *Plant Physiology*, 162(1), 256–267. https://doi.org/10.1104/pp.113.216721

FAOSTAT. (2023). *Food and Agriculture Organization of the United Nations*. https://www.fao.org/faostat/

Fracheboud, Y., Haldimann, P., Leipner, J., & Stamp, P. (2004). Chlorophyll fluorescence as a selection tool for cold tolerance of photosynthesis in maize (*Zea mays* L.). *Journal of Experimental Botany*, 55(396), 1187–1194. https://doi.org/10.1093/jxb/erh129

Grobman, A., Bonavia, D., Dillehay, T. D., Piperno, D. R., Iriarte, J., & Holst, I. (2012). Preceramic maize from Paredones and Huaca Prieta, Peru. *PNAS*, 109(5), 1755–1759. https://doi.org/10.1073/pnas.1120270109

Guo, L., Wang, X., Zhao, M., Huang, C., Li, C., Li, D., ... & Tian, F. (2018). Stepwise cis-regulatory changes in ZCN8 contribute to maize flowering-time adaptation. *Current Biology*, 28(18), 3005–3015.

Hammer, G. L., Dong, Z., McLean, G., Doherty, A., Messina, C., Schussler, J., & Cooper, M. (2009). Can changes in canopy and/or root system architecture explain historical maize yield trends in the US corn belt? *Crop Science*, 49(1), 299–312. https://doi.org/10.2135/cropsci2008.03.0152

Hallauer, A. R., Carena, M. J., & Miranda Filho, J. D. (2010). *Quantitative genetics in maize breeding* (Vol. 6). Springer Science & Business Media.

Hatfield, J. L., & Prueger, J. H. (2015). Temperature extremes: Effect on plant growth and development. *Weather and Climate Extremes*, 10, 4–10. https://doi.org/10.1016/j.wace.2015.08.001

Hufford, M. B., Xu, X., van Heerwaarden, J., Pyhäjärvi, T., Chia, J. M., Cartwright, R. A., & Ross-Ibarra, J. (2012). Comparative population genomics of maize domestication and improvement. *Nature Genetics*, 44(7), 808–811. https://doi.org/10.1038/ng.2309

Hung, H. Y., Shannon, L. M., Tian, F., Bradbury, P. J., Chen, C., Flint-Garcia, S. A., McMullen, M. D., Ware, D., Buckler, E. S., Doebley, J. F., et al. (2012) *ZmCCT* and the genetic basis of day-length adaptation underlying the postdomestication spread of maize. *Proceedings of the National Academy of Science USA*, 109: E1913–1921

Kistler, L., Maezumi, S. Y., de Souza, J. G., Przelomska, N. A. S., Costa, F. M., Smith, O., & Brown, T. A. (2018). Multiproxy evidence highlights a complex evolutionary legacy of maize in South America. *Science*, 362(6420), 1309–1313. https://doi.org/10.1126/science.aav0207

Leng, P., Zhao, J., Li, H., Wang, G., & Liu, H. (2014). The Maize DREB transcription factor ZmDREB1A mediates cold stress responses in transgenic Arabidopsis. *PLOS ONE*, 9(10), e109687. https://doi.org/10.1371/journal.pone.0109687

Li, H., Peng, Z., Yang, X., Wang, W., Fu, J., Wang, J., & Xu, Y. (2013). Genome-wide association study dissects the genetic architecture of oil biosynthesis in maize kernels. *Nature Genetics*, 45(1), 43–50. https://doi.org/10.1038/ng.2484

Lobell, D. B., Schlenker, W., & Costa-Roberts, J. (2011). Climate trends and global crop production since 1980. *Science*, 333(6042), 616–620. https://doi.org/10.1126/science.1204531

Goering, R., Larsen, S., Tan, J., Whelan, J., & Makarevitch, I. (2021). QTL mapping of seedling tolerance to exposure to low temperature in the maize IBM RIL population. *Plos One*, 16(7), e0254437.

Matsuoka, Y., Vigouroux, Y., Goodman, M. M., Sanchez, G. J., Buckler, E., & Doebley, J. (2002). A single domestication for maize shown by multilocus microsatellite genotyping. *Proceedings of the National Academy of Sciences*, 99(9), 6080–6084. https://doi.org/10.1073/pnas.052125199

Meihls, L. N., Kaur, H., & Jander, G. (2012). Natural variation in maize defense against insect herbivory. *Current Opinion in Plant Biology*, 15(4), 470–476. https://doi.org/10.1016/j.pbi.2012.05.004

Mercer, K. L., & Perales, H. R. (2010). Evolutionary response of landraces to climate change in Mexico. *Global Environmental Change*, 20(1), 49–57. https://doi.org/10.1016/j.gloenvcha.2009.06.001

Merrill, W. L., Hard, R. J., Mabry, J. B., Fritz, G. J., Adams, K. R., Roney, J. R., & MacWilliams, A. C. (2009). The diffusion of maize to the southwestern United States and its impact. *Proceedings of the National Academy of Sciences*, 106(50), 21019–21026.

Meng, X., Muszynski, M. G., & Danilevskaya, O. N. (2011). The FT-like ZCN8 gene functions as a floral activator and is involved in photoperiod sensitivity in maize. *The Plant Cell*, 23(3), 942–960.

Mir, C., Zerjal, T., Combes, V., Dumas, F., Madur, D., Bedoya, C., & Charcosset, A. (2013). Out of America: Tracing the genetic footprints of the global diffusion of maize. *Theoretical and Applied Genetics*, 126(11), 2671–2682. https://doi.org/10.1007/s00122-013-2164-z

Monfreda, C., Ramankutty, N., & Foley, J. A. (2008). Farming the planet: 2. Geographic distribution of crop areas, yields, physiological types, and net primary production in the year 2000. *Global Biogeochemical Cycles*, 22(1). https://doi.org/10.1029/2007GB002947

Morris, M. L. (Ed.). (1998). *Maize seed industries in developing countries*. Lynne Rienner Publishers.

Oerke, E. C. (2006). Crop losses to pests. *The Journal of Agricultural Science*, 144(1), 31–43. https://doi.org/10.1017/S0021859605005708

Pressoir, G., & Berthaud, J. (2004). Population structure and strong divergent selection shape phenotypic diversification in maize landraces. *Heredity*, 92(2), 95–101.

Piperno, D. R., & Flannery, K. V. (2001). The earliest archaeological maize (*Zea mays* L.) from highland Mexico: New accelerator mass spectrometry dates and their implications. *PNAS*, 98(4), 2101–2103. https://doi.org/10.1073/pnas.98.4.2101

Piperno, D. R., Ranere, A. J., Holst, I., Iriarte, J., & Dickau, R. (2009). Starch grain and phytolith evidence for early ninth millennium B.P. maize from the Central Balsas River Valley, Mexico. *Proceedings of the National Academy of Sciences*, 106(13), 5019–5024. https://doi.org/10.1073/pnas.0812590106

Qin, F., Kakimoto, M., Sakuma, Y., Maruyama, K., Osakabe, Y., Tran, L. S. P., & Yamaguchi-Shinozaki, K. (2011). Regulation and functional analysis of *ZmDREB2A* in maize under drought and heat stresses. *Plant Journal*, 66(4), 605–617. https://doi.org/10.1111/j.1365-313X.2011.04513.x

Radin, P. (1991). *The road of life and death: a ritual drama of the American Indians* (Vol. 5). Princeton University Press.

Revilla, P., Rodríguez, V. M., Ordás, B., Rincent, R., Charcosset, A., Giauffret, C., & Malvar, R. A. (2016). Cold tolerance in maize. *Frontiers in Plant Science*, 7, 1479. https://doi.org/10.3389/fpls.2016.01479

Riva-Roveda, L., Escale, B., Giauffret, C., & Périlleux, C. (2016). Maize plants can enter a standby mode to cope with chilling stress. *BMC Plant Biology*, 16, 212. https://doi.org/10.1186/s12870-016-0909-y

Romay, M. C., Millard, M. J., Glaubitz, J. C., Peiffer, J. A., Swarts, K., Casstevens, T. M., & Buckler, E. S. (2013). Comprehensive genotyping of the USA national maize inbred seed bank. *Genome Biology*, 14(6), R55. https://doi.org/10.1186/gb-2013-14-6-r55

Roth, B. J., Freeman, A. K. L., & Wills, W. H. (2018). Maize and cultural complexity: Stable isotopes, diet, and early agriculture in the American Southwest. *Journal of Anthropological Archaeology*, 49, 1–13. https://doi.org/10.1016/j.jaa.2017.12.003

Salvi, S., Sponza, G., Morgante, M., Tomes, D., Niu, X., Fengler, K., & Tuberosa, R. (2007). Conserved noncoding genomic sequences associated with a flowering-time quantitative trait locus in maize. *PNAS*, 104(27), 11376–11381. https://doi.org/10.1073/pnas.0704145104

Schnable, P. S., Ware, D., Fulton, R. S., Stein, J. C., Wei, F., Pasternak, S., ... & Presting, G. G. (2009). The B73 maize genome: complexity, diversity, and dynamics. *Science*, 326(5956), 1112–1115.

Studer, A., Zhao, Q., Ross-Ibarra, J., & Doebley, J. (2011). Identification of a functional transposon insertion in the maize domestication gene *tb1*. *Nature Genetics*, 43(11), 1160–1163. https://doi.org/10.1038/ng.942

Swarts, K., Gutaker, R. M., Benz, B., Blake, M., Bukowski, R., Holland, J., & Ross-Ibarra, J. (2017). Genomic estimation of complex traits reveals ancient maize adaptation to temperate North America. *Science*, 357(6350), 512–515. https://doi.org/10.1126/science.aam9425

Thornsberry, J. M., Goodman, M. M., Doebley, J., Kresovich, S., Nielsen, D., & Buckler, E. S. (2001). Dwarf8 polymorphisms associate with variation in flowering time. *Nature genetics*, 28(3), 286–289.

Troyer, A. F. (2004). Background of US hybrid corn II: Breeding, climate, and food. *Crop Science*, 44(2), 370–380.

Tenaillon, M. I., & Charcosset, A. (2011). A European perspective on maize history. *Comptes Rendus Biologies*, 334(3), 221–228. https://doi.org/10.1016/j.crvi.2010.12.016

Yang, Q., Li, Z., Li, W., Ku, L., Wang, C., Ye, J., ... & Xu, M. (2013). CACTA-like transposable element in ZmCCT attenuated photoperiod sensitivity and accelerated the post-domestication spread of maize. *Proceedings of the National Academy of Sciences*, 110(42), 16969–16974.

Van Heerwaarden, J., Hufford, M. B., & Ross-Ibarra, J. (2011). Genetic signals of origin, spread, and introgression in a large sample of maize landraces. *PNAS*, 108(3), 1088–1092. https://doi.org/10.1073/pnas.1013011108

Varshney, R. K., Bohra, A., Yu, J., Graner, A., Zhang, Q., & Sorrells, M. E. (2021). Designing future crops: Genomics-assisted breeding comes of age. *Trends in Plant Science*, 26(6), 631–649. https://doi.org/10.1016/j.tplants.2021.03.010

Xu, Y., Liu, X., Fu, J., Wang, H., Wang, J., & Huang, C. (2020). Enhancing genetic gain through genomic selection: From livestock to plants. *Plant Communications*, 1(3), 100005. https://doi.org/10.1016/j.xplc.2019.100005

Zhao, C., Liu, B., Piao, S., Wang, X., Lobell, D. B., Huang, Y., & Asseng, S. (2017). Temperature increase reduces global yields of major crops in four independent estimates. *Proceedings of the National Academy of Sciences*, 114(35), 9326–9331. https://doi.org/10.1073/pnas.1701762114

Zhao, D., Reddy, K. R., Kakani, V. G., Read, J. J., & Koti, S. (2006). Nitrogen deficiency effects on plant growth, leaf photosynthesis, and hyperspectral reflectance properties of sorghum. *European Journal of Agronomy*, 24(4), 472–478. https://doi.org/10.1016/j.eja.2006.01.007

4 Abiotic Stresses and Maize

4.1 INTRODUCTION

Crops form the cornerstone of global food systems, supplying nearly 80% of the human diet. Among these, cereals alone contribute approximately 50% to the total food production worldwide (Langridge & Fleury, 2011). Maize (*Zea mays*), commonly known as corn—particularly in the United States—has emerged as one of the most critical cereal crops, with a global production surpassing 1 billion tons since 2013 (FAOSTAT, 2017). Its versatility enables it to meet growing dietary demands both directly as a staple food and indirectly as a major component of livestock feed. Beyond its nutritional value, maize also serves as a renewable energy source through bioethanol production, making it a strategic crop in the quest for sustainable energy alternatives (Vaughan et al., 2018). However, the sustainability and productivity of maize are increasingly threatened by the adverse effects of climate change. Fluctuations in weather patterns, especially extreme temperatures, drought, nutrient imbalances, and salinity, have emerged as key abiotic stressors that significantly hinder maize growth and yield. Among these, temperature and precipitation stand out as the most influential climatic factors, with solar radiation also playing a critical role in determining crop performance (Xu et al., 2016). Current climate anomalies such as prolonged droughts, heatwaves, and waterlogging have led to considerable yield losses. For instance, annual maize production losses due to drought and waterlogging have been estimated at 20–30% in China and 25–30% in India (Zaidi et al., 2010; Ahuja et al., 2010).

Despite the central role of crops in global food supply, the existing agricultural productivity lags behind the ever-escalating food demand. Projections suggest that by 2050, the global population will increase by approximately 2 billion, placing unprecedented pressure on food systems (Shelden & Roessner, 2013). This anticipated scenario is further complicated by a decline in arable land, making it imperative to enhance crop yields and resilience. To bridge the gap between food demand and supply, developing climate-resilient, high-yielding maize varieties is essential. This goal can be achieved through an integrated approach that combines traditional breeding techniques with modern molecular tools, including transgenic technologies (Cairns et al., 2013). The implementation of globally coordinated strategies aimed at improving per-hectare productivity and engineering crops to withstand environmental extremes is of paramount importance. Maize, due to its agronomic significance and genetic plasticity, stands out as a prime candidate for such genetic advancements (Vinocur & Altman, 2005).

In this context, maize varieties with enhanced tolerance to abiotic stresses—particularly drought and heat—are indispensable for ensuring food security in the face of

DOI: 10.1201/9781003634027-4

climate change (Fedoroff et al., 2010). This chapter aims to explore the multifaceted response mechanisms of maize to various abiotic stresses, including drought, low temperature (LT), salinity stress, heat stress, and heavy metal stress, encompassing morphological, physiological, biochemical, and molecular dimensions. Furthermore, it outlines the prospective challenges and strategic directions necessary for developing climate-smart maize germplasm capable of sustaining global food demands amid mounting environmental uncertainties.

4.2 DROUGHT STRESS TOLERANCE IN MAIZE: PRESENT STATUS AND FUTURE PROSPECTS

Drought is a significant global challenge, severely impacting crop production, food security, and human well-being. As one of the most critical abiotic stresses, it affects nearly a third of the world's population and has led to substantial agricultural losses. Between 2005 and 2015, drought caused an estimated USD 29 billion in direct damages to agriculture in developing countries (FAO, 2021). This ongoing threat diminishes the effectiveness of modern agricultural advancements, including fertilization and technological innovations (Lobell et al., 2011). Therefore, developing high-yielding crop varieties with enhanced drought tolerance is crucial (Fita et al., 2015; Raza et al., 2019). Maize (*Zea mays* L.), a major biofuel and biotech crop—92% of which is genetically modified in the United States—has gained prominence due to its diverse applications beyond food and feed (Schwietzke et al., 2009; Yadava et al., 2017). Drought is a major limiting factor in maize production, particularly as nearly 80% of wet-season maize is cultivated under rainfed conditions, while dry-season maize faces severe water shortages (Trachsel et al., 2016). The extent of yield losses varies between 30% and 90%, depending on the severity and duration of drought stress, with the flowering and grain-filling stages being most affected (Sah et al., 2020). In many regions, maize is grown in rainfed systems with only 300–500 mm of precipitation—insufficient to achieve optimal yields (Sah et al., 2020). Given that genetic improvements alone are expected to close only about 30% of the yield gap under water stress, a deeper understanding of maize genetics and its application in breeding for drought tolerance is essential for stabilizing global production. Advanced molecular breeding techniques, including marker-assisted selection (MAS), transgenics, and genome editing, offer promising solutions to enhance maize resilience against drought stress. Recent projections indicate that by 2050, the global average temperature could rise by more than 2°C under the current high-emission scenario, leading to an estimated annual maize yield loss of 10 million tons due to rising temperatures and shifting rainfall patterns (Tesfaye et al., 2018). In response, modern molecular approaches in the genomics era have focused on identifying and manipulating drought-related traits, genes, and quantitative trait loci (QTLs) to enhance maize resilience (Deikman et al., 2012; Kumar et al., 2022). Additionally, various omics platforms enable the comprehensive analysis of transcriptomes, proteomes, and metabolomes involved in drought tolerance mechanisms (Yu et al., 2021). This section of the chapter aims to evaluate recent advancements and the potential benefits of novel breeding technologies in developing drought-tolerant maize. Understanding the molecular and physiological mechanisms underlying drought tolerance is crucial for integrating advanced breeding tools with biotechnology to improve crop resilience.

4.2.1 Morpho-Physiological Responses of Maize to Drought

Drought stress has a profound and multifaceted impact on plant systems, triggering complex physiological and cellular responses (Figure 4.1).

In maize, the vegetative, silking (flowering), and ear (grain-filling) stages are particularly vulnerable, with yield losses of approximately 25%, 50%, and 21%, respectively (Sah et al., 2020). Drought disrupts key molecular and physiological processes, including photosynthesis, abscisic acid (ABA) accumulation, osmotic adjustment, and antioxidant defense mechanisms (Gong et al., 2014). In response to water deficit, adaptive traits such as osmotic adjustment, dehydration tolerance, and reduced photosynthetic activity are activated. A decline in photosynthesis leads to stomatal closure and decreased production of photosynthetic enzymes. At the cellular level, drought stress induces turgor loss and alters membrane fluidity and composition. Increased ABA levels during drought initiate a signal transduction cascade that lowers the osmotic potential of guard cells by promoting the loss of K^+ and Cl^-, ultimately causing stomatal closure (Anami et al., 2009). Morphologically, drought stress disrupts plant growth and development by interfering with cell cycle regulation (Carneiro et al., 2021). The perception of abiotic stress triggers signaling cascades that activate cell cycle checkpoints, leading to impaired G1-to-S phase

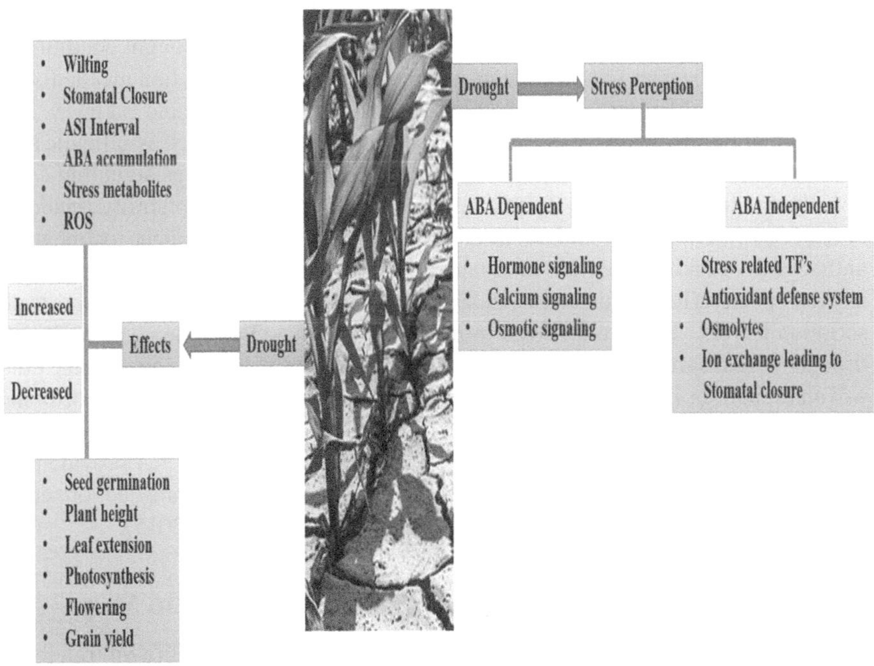

FIGURE 4.1 Impact of drought stress on maize and the stress perception.

transition, slowed DNA replication, and delayed mitotic entry (Kadota et al., 2005; Qi & Zhang, 2020). Water stress also shortens meristematic regions in wheat and maize leaves and prolongs the cell cycle due to reduced cyclin-dependent kinase (CDK) activity (Granier et al., 2000). Adaptations in root and shoot growth, such as an increased root-to-shoot ratio, enhanced tissue water storage capacity, thicker cuticles, and altered water permeability, play a crucial role in drought tolerance, with root growth modifications being particularly vital for crop resilience (Verslues et al., 2006). Under severe water stress, young maize plants respond by rolling their leaves, while reproductive development is significantly affected, leading to delayed spikelet formation, an extended anthesis-silking interval (ASI), silk senescence, and pollen abortion (Lu et al., 2011; Stomatal closure is one of the earliest responses to drought stress, reducing transpiration and CO_2 absorption, which, in turn, lowers photosynthetic activity (Wang et al., 2018). Severe drought during the flowering stage can lead to complete ear abortion, rendering the plant barren. Additionally, drought limits leaf expansion, plant height, and tassel development, all of which contribute to reduced crop yield by restricting the availability of photosynthates needed for proper growth (Aslam et al., 2015).

To cope with water scarcity, many plants, including maize, reallocate growth and dry matter accumulation from the shoot to the root system (Maazou et al., 2016). At the metabolic level, transient increases in amino acid concentrations occur in the sap of water-stressed (WS) plants, with sustained accumulation only under severe stress. Changes in solute concentrations, along with modifications in protein–protein and protein–lipid interactions, are key metabolic adjustments linked to drought stress (Valliyodan & Nguyen, 2006). Additionally, the production of osmoprotectants and osmolytes plays a crucial role in the plant's defense response to drought conditions.

In response to water scarcity, plants have evolved various adaptation mechanisms, including phytoalexin production, activation of the phenylpropanoid pathway, and induction of lignin biosynthesis. Stress-related small molecules such as salicylic acid (SA), methyl salicylate, jasmonic acid, and methyl jasmonate act as signaling compounds that trigger systemic defense responses and enhance adaptability (Nadeem et al., 2019). Meanwhile, antioxidants like ascorbic acid, glutathione, tocopherols, anthocyanins, and carotenoids mitigate oxidative damage by scavenging reactive oxygen species (ROS) generated under stress conditions. Several genes and transcription factors (TFs) are involved in modulating plant cellular metabolism during drought stress, either by maintaining osmotic homeostasis, facilitating damage control and repair, or regulating gene expression. Cellular ionic, osmotic, and hormonal (ABA) signaling pathways activate numerous stress-responsive genes that not only regulate gene expression patterns but also enhance osmotolerance and provide protection. Additionally, these pathways play a crucial role in signal transduction, coordinating the plant's overall response to drought stress (Lata et al., 2015).

4.2.2 Approaches to Study the Drought Stress Tolerance to Maize

Advancements in high-throughput genotyping and the availability of reference genome datasets have significantly enhanced the study of economically important traits. Two primary approaches—linkage mapping and association mapping—are widely used to identify complex traits associated with drought tolerance. However, pinpointing candidate genes remains a challenge, prompting the development of advanced methods such as Mendelian randomization (MR) analysis and transcriptome-wide association studies (TWAS). These approaches help assess expression-trait associations and identify candidate genes based on gene expression variation, improving the precision of trait mapping (Zhu et al., 2016).

4.2.2.1 Quantitative Trait Locus (QTL) through Linkage Mapping

The identification of QTLs is a key strategy for enhancing drought tolerance in maize through MAS. Over the past decades, numerous studies have focused on mapping genes and QTLs associated with drought resilience in maize (Ribaut et al., 1997; Tuberosa et al., 2002; Lu et al., 2006; Liu et al., 2011; Rahman et al., 2011; Zhu et al., 2011; Zhao et al., 2018). Advances in sequencing technologies have facilitated the development of ultra-high-density genetic maps, enabling the discovery of extensive molecular markers and precise localization of QTLs. Several studies have reported QTLs linked to distinct phenotypic traits in maize under drought stress across various mapping populations.

Frova et al. (1999) identified QTLs associated with flowering time and plant height under both well-watered (WW) and WS conditions. Nikolić et al. (2011) mapped 43 QTLs related to grain yield (GY) and morphological traits across all maize chromosomes except chromosome 9. Among these, five QTLs for GY accounted for 0.1–15.86% of the phenotypic variation. Additionally, 12 QTLs were linked to leaf width, 15 to plant height and ear height, and one to leaf number on chromosome 6. Almeida et al. (2013) identified 83 and 62 genomic regions controlling GY and ASI, respectively, across multiple environments and genetic backgrounds. Zhao et al. (2018) reported 69 QTLs for plant height, ear height, ASI, ear weight, cob weight, 100-kernel weight, and ear length under both drought-stressed and control conditions, explaining 4.0–17.2% of the observed phenotypic variation.

Linkage mapping is a widely used approach for locating QTLs; however, it typically identifies only a limited number of QTLs. Additionally, fine mapping and cloning of genes within QTL regions require large populations to achieve high-resolution genetic maps (Dinka et al., 2007), making the process resource-intensive and time-consuming. The presence of repetitive sequences, which constitute approximately 85% of the maize genome (Schnable et al., 2009), further complicates and slows down QTL fine mapping and gene cloning.

Although extensive genetic studies have dissected drought tolerance in maize, the successful application of identified QTLs in breeding programs remains limited. This challenge arises due to the lack of QTL validation across diverse genetic and environmental backgrounds, genetic complexity, epistasis, significant QTL × environment interactions (QEI), population-specific QTLs, and the use of agronomically inferior donor lines (Vargas et al., 2006; Liu et al., 2011).

To overcome these limitations, meta-QTL (MQTL) analysis has emerged as a powerful tool for identifying stable and significant QTLs across different populations and environmental conditions. This approach facilitates the discovery of genomic regions associated with GY under both WS and WW conditions across various genotypes (Swamy et al., 2011). Zhao et al. (2018) identified 36 MQTLs in 26 populations under WW and WS conditions, along with 39 candidate genes within these regions. Similarly, Almeida et al. (2013) employed MQTL analysis to detect seven genomic regions associated with GY and one linked to ASI on chromosomes 1, 4, 5, 7, and 10, which were consistently expressed under both WS and WW conditions.

4.2.2.2 Genome-Wide Association Studies

Traditional breeding methods have captured only limited allelic variation for drought tolerance traits. To overcome this limitation, genome-wide association studies (GWAS) have been widely adopted to identify favorable alleles associated with drought resilience in maize. Unlike linkage mapping, which relies on controlled populations, GWAS leverages historical recombination events in diverse natural populations, enabling higher-resolution QTL fine mapping (Schaid et al., 2018). GWAS were first applied in maize as a candidate gene association study at the beginning of the 21st century (Thornsberry et al., 2001; Beló et al., 2008). Since the release of the maize reference genome (Schnable et al., 2009), GWAS have become a powerful tool for dissecting numerous economically important traits under drought stress

Recent advancements highlight GWAS as an effective approach for associating genotypic and phenotypic variations. Additionally, it facilitates the mapping of expression QTLs (eQTLs) by leveraging transcriptomic data, providing insights into the genetic basis of gene expression traits and explaining a significant proportion of phenotypic variation. For instance, Fu et al. (2013) analyzed the expression of 14,375 genes, with each eQTL explaining, on average, more than 15% of the phenotypic variance per trait. However, drought tolerance is a highly complex trait influenced by multiple genetic and environmental factors, making it challenging to dissect through GWAS alone. To address this, researchers often study correlated metabolic traits or seedling survival rates under drought stress to identify candidate genes involved in drought adaptation (Setter et al., 2011; Liu et al., 2013; Farfan et al., 2015; Mao et al., 2015; Zhang et al., 2016). Several association studies have successfully identified loci linked to drought tolerance traits, such as carbohydrate and ABA metabolite accumulation (Setter et al., 2011), dehydration-responsive element-binding (DREB) genes, and other stress-related genes (Liu et al., 2013; Thirunavukkarasu et al., 2014; Wallace et al., 2016). Wang et al. (2016) further identified 83 genetic variants for drought resistance and pinpointed the ZmVPP1 gene, which has been demonstrated to enhance drought tolerance in maize.

4.2.2.3 Omics Approaches

Recent advancements in plant omics, including genomics, epigenomics, transcriptomics, proteomics, and metabolomics, have significantly enhanced our understanding of the complex mechanisms underlying drought stress tolerance (Budak et al., 2015).

Drought tolerance is a highly intricate trait regulated at multiple levels, including gene expression, post-translational modifications, and protein interactions. Omics-based studies provide valuable insights into these regulatory mechanisms, aiding in the development of drought-tolerant crop cultivars (Wu et al., 2017). Omics approaches have been extensively used to identify key genes, QTLs, gene networks, and regulatory pathways associated with drought tolerance in maize (Table 4.1).

Genomic tools allow the identification of gene functions and their interactions within regulatory networks, while transcriptomics, proteomics, and metabolomics

TABLE 4.1
Effects of Drought Stress and Adaptive Responses in Maize

Biological Level	Drought Impact	Adaptive/Defense Responses
Growth Stages Affected	Vegetative (25% yield loss), Silking (50%), Grain filling (21%) (Sah et al., 2020)	Timing and stage-specific management essential for resilience
Morphological	Reduced leaf expansion, plant height, tassel development; increased leaf rolling; delayed spikelet formation; extended ASI; ear and pollen abortion	Altered root-to-shoot ratio, thicker cuticle, modified water permeability, enhanced root growth
Physiological	Decreased photosynthesis, stomatal closure, ABA accumulation, reduced enzyme activity	Osmotic adjustment, enhanced water-use efficiency, dehydration tolerance
Biochemical/ Metabolic	Membrane disruption, turgor loss, reduced CDK activity, protein–protein/lipid interactions altered	Accumulation of osmoprotectants, osmolytes, and amino acids; lignin biosynthesis
Phytochemical	ROS overproduction, oxidative stress	Increased antioxidant activity (ascorbic acid, glutathione, tocopherols, anthocyanins, carotenoids)
Molecular/Genetic	Stress-induced gene expression changes; altered signal transduction pathways	ABA, osmotic, and ionic signaling cascades; activation of drought-responsive genes and TFs
Cellular/Cell Cycle	Impaired G1–S transition, slowed replication, delayed mitosis; shortened meristem length	Stress signal perception and checkpoint regulation; possible recovery under moderate stress
Hormonal & Signal Molecules	ABA elevation, altered hormonal balance	Production of salicylic acid, jasmonic acid, methyl derivatives; activation of systemic defenses
Yield Components	Reduced photosynthate availability; ear and grain abortion	Growth reallocation to reproductive tissues; selection for tolerant hybrids

facilitate the mapping of drought-responsive traits onto segregating populations. By integrating these omics technologies, researchers have gained a comprehensive view of maize responses to drought at different developmental stages and in various tissues, including seeds, leaves, and roots. For instance, Huang et al. (2012) investigated maize seed desiccation tolerance and identified differentially expressed proteins, such as HSP3, EMB564, and stress response proteins, that may contribute to drought resilience during embryogenesis and germination. Similarly, to better understand how ABA modulates the maize proteome under drought stress, Hu et al. (2011, 2012) conducted a comparative proteomic analysis between an ABA-deficient maize mutant (*vp5*) and its wild-type counterpart (*Vp5*), revealing key proteins involved in drought adaptation.

Furthermore, several QTLs and genes associated with kernel desiccation have been linked to ABA biosynthesis through QTL mapping (Capelle et al., 2010) and oligo-microarray analyses (Luo et al., 2010). Many drought tolerance-related genes and proteins, such as ZmTPA (Huang et al., 2012), ZmRFP1 (Xia et al., 2012), and ZmCPK4

(Jiang et al., 2013), have been found to be induced by ABA or regulated in an ABA-dependent manner. Bonhomme et al. (2012) identified 3,664 unique phosphorylation sites on 2,496 proteins that influence multiple biological processes, including epigenetic regulation, transcriptional control, phytohormone signaling (ABA, ethylene (ET), auxin, jasmonate), and cell cycle regulation under drought stress.

Proteomic studies have further contributed to understanding drought-induced changes in maize. Benešová et al. (2012) examined drought-responsive alterations in the maize leaf proteome, while Zhang et al. (2021) identified the Bx12 and ZmGLK44 genes as key regulators of metabolite biosynthesis and drought tolerance. Gaffney et al. (2021) reported several drought-related biomarkers, including neophaseic acid, hydroxyabscisic acid, methyl itaconate, phospholipids, and lysolecithin.

Omics studies have revealed that maize drought responses involve several key protein groups, including:

- **Protective proteins**, such as heat shock proteins (HSPs) (Hu et al., 2010; Benešová et al., 2012; Wang et al., 2019)
- **Stress response-related proteins**, like NBS-LRR resistance-like proteins (Hu et al., 2011)
- **14-3-3-like proteins** (Li et al., 2009; Huang et al., 2012)
- **Late embryogenesis abundant (LEA) proteins** (Huang et al., 2012; Wang et al., 2019)
- **Phytohormone-related and signaling proteins**, such as auxin-repressed proteins and serine/threonine protein kinases (Luo et al., 2010; Hu et al., 2011; Bonhomme et al., 2012; Dong et al., 2020; Li et al., 2021)

Despite the vast amount of omics data generated, a significant limitation is that most studies have been conducted under controlled laboratory conditions, which often fail to capture the complexity of field environments. There are substantial differences in protein markers involved in similar metabolic pathways under lab versus field conditions. Additionally, many drought stress experiments focus on acute responses without considering long-term adaptation, injury, and recovery mechanisms (Lyon et al., 2016). Consequently, while these studies expand the catalog of drought-responsive proteins, their direct application to crop improvement remains limited. A major challenge in translating omics discoveries into practical breeding applications is the large-scale genotype-by-environment (G × E) interactions, which complicate the identification of universally effective drought tolerance mechanisms. Therefore, integrating omics research with field-based trials is crucial to uncovering the systematic mechanisms of drought responses and applying these findings to improve drought resilience in maize. Future research should prioritize field validation of omics data to bridge the gap between laboratory findings and real-world agricultural applications.

4.2.3 Epigenetic Modifications to Improve Drought Tolerance

Epigenetics has emerged as a crucial research area for understanding biochemical modifications in DNA that regulate gene expression without altering the DNA sequence. These modifications are particularly significant in response to environmental stresses such as drought and can be long-lasting, heritable, and contribute to plant stress memory.

Key epigenetic mechanisms—including small RNAs (miRNAs and siRNAs) for post-transcriptional gene silencing, DNA methylation, and histone modifications—play a vital role in modulating gene activity under drought stress (Thiebaut et al., 2019; Wojtyla et al., 2020). These processes help plants enhance stress resilience by dynamically regulating gene expression. Specifically, small RNAs (sRNAs) contribute to drought tolerance by:

- Downregulating negative regulators of stress responses
- Upregulating positive regulators involved in stress adaptation
- Modulating plant hormones, ROS levels, and TFs (Banerjee et al., 2017)

By fine-tuning these molecular pathways, epigenetic modifications provide plants with a flexible and adaptive mechanism to cope with environmental challenges, including drought stress.

Significant advancements have been made in quantifying epigenetic variations and understanding their impact on plant growth, development, yield, and quality (Cortijo et al., 2014). Epigenetic regulation involves various post-transcriptional and post-translational modifications, including phosphorylation, ubiquitination, sumoylation, DNA methylation, RNA interference, histone modifications, functional proteins, and chromatin remodeling (Thiebaut et al., 2019). By integrating traditional breeding principles, selecting for epigenetic phenotypes can enhance the development of stress-tolerant plants at multiple levels—ranging from metabolite production to polygenic trait expression. Regulatory and epigenetic factors play a crucial role in gene expression and the selection of genotypes with improved yield and quality (Verkest et al., 2015).

Epigenetic modifications can dynamically regulate gene transcription in response to environmental stimuli, making them essential for stress adaptation and crop resilience. Mechanistic studies of these modifications will provide deeper insights into the specific genes and regulatory regions involved in abiotic stress responses. This knowledge can help identify target pathways for crop improvement, ultimately contributing to the development of more resilient and high-yielding plant varieties (Sun et al., 2021). DNA methylation is a crucial epigenetic modification associated with plant development and stress responses (Huang et al., 2019). This process is regulated by DNA methyltransferase enzymes, which are influenced by various plant regulatory pathways to facilitate rapid and adaptive responses to biotic and abiotic stresses (Zhang et al., 2018). Wang et al. (2021) used MeDIP-Seq to map DNA methylation patterns in four maize inbred lines, revealing distinct drought stress responses between drought-tolerant and drought-sensitive genotypes. The study found that DNA methylation not only suppresses gene expression but also influences alternative splicing in response to drought stress. Under water-limited conditions, methylation levels increased, and the methylome of drought-tolerant inbred lines remained more stable than that of drought-sensitive lines. These findings provide valuable insights into the role of DNA methylation in drought tolerance, highlighting its potential involvement in gene regulation and alternative splicing mechanisms.

4.2.4 Breeding for Drought Stress Tolerance in Maize

4.2.4.1 Conventional Breeding

Improving drought tolerance through conventional breeding has been a significant challenge due to the complex genetic nature of the trait. However, several drought-tolerant maize cultivars—such as Oh605, ND2005, ND2006, and 16 tropically adapted inbred lines (TZEI 1 to TZEI 16), as well as TZE-W Pop DT STR C4 and TZE-Y Pop DT STR C4—have been successfully developed using traditional breeding methods like recurrent selection, inbreeding, pedigree breeding, backcrossing, and hybridization (Ashraf, 2010). The success of conventional breeding depends largely on the availability of diverse germplasm and the identification of suitable donor lines carrying drought-tolerant genes. Several drought-tolerant germplasm sources have been identified in maize (Cairns et al., 2013). However, this approach is often time-consuming, labor-intensive, and costly, requiring multiple selection and breeding cycles. Additionally, conventional breeding may introduce undesirable genes due to linkage drag, further complicating the process. Moreover, drought tolerance is a polygenic trait with strong interactions with yield-related characteristics, making its improvement through traditional breeding methods less efficient (Ahmar et al., 2020). As a result, advancements in next-generation sequencing (NGS) and molecular breeding techniques have paved the way for more precise and efficient approaches to developing drought-tolerant maize.

4.2.4.2 Molecular Breeding

The application of molecular markers has significantly enhanced the efficiency of breeding drought-tolerant crops (Andjelković & Ignjatović-Micić, 2011; Mwamahonje et al., 2021). Unlike conventional breeding, genomics allows for the dissection of quantitative traits into specific genetic determinants, enabling MAS and ultimately facilitating QTL cloning and genetic engineering (Tuberosa & Salvi, 2006; Oladosu et al., 2019). Genomics-assisted breeding techniques such as MAS, marker-assisted backcrossing (MABC), marker-assisted recurrent selection (MARS), and genomic selection (GS) have opened new avenues for improving drought tolerance. Under WS conditions, Ribaut and Ragot (2007) successfully used MABC to introduce five QTLs associated with GY and flowering traits into a drought-prone maize line, resulting in higher GY and reduced ASI. However, MABC primarily focuses on large-effect QTLs that have been identified and validated across diverse genetic backgrounds before being introduced into elite varieties (Bernardo, 2008). This ensures that the selected traits contribute significantly and consistently to drought tolerance without undesirable genetic trade-offs. Unlike MABC, MARS utilizes a subset of markers significantly associated with target traits to accumulate multiple medium-effect QTLs within a given population (Bernardo, 2008; Ribaut et al., 2010). Beyene et al. (2016) demonstrated the potential of MARS in increasing genetic gain for GY under WS conditions by studying 10 biparental maize populations. Another marker-based approach, GS, leverages all available markers, regardless of their significance, to predict the genetic value of progenies for selection (Lorenz, 2013). Using GS, Beyene et al. (2015) evaluated genetic gains in GY across eight biparental maize populations under managed drought stress conditions. The study revealed that GS is more effective than traditional pedigree-based phenotypic selection for improving genetic gains in tropical maize under drought stress. Through a "breeding by design" strategy, breeders can combine identified QTLs and transgenes relevant to yield and drought tolerance traits into elite genotypes. By analyzing allelic variations and the genetic basis of secondary traits, new ideotype crops can be designed with enhanced drought resilience.

Another key application of molecular breeding is gene cloning, particularly for genes linked to drought-tolerant QTLs. Various techniques have been employed to map and clone candidate genes associated with stress responses (Salvi & Tuberosa, 2005; Tondelli et al., 2006). For instance, Masle et al. (2005) successfully cloned the ERECTA gene in Arabidopsis thaliana, which influences transpiration efficiency. However, no reports exist on the successful cloning of genes underlying QTLs in crop species, as identifying a suitable QTL for gene cloning remains a challenge (Hu et al., 2021). Despite significant advancements, molecular breeding research has primarily focused on detecting drought-responsive traits, with limited progress in transferring QTLs into elite high-yielding genotypes. While MAS holds promise, its contribution to developing drought-tolerant cultivars has been modest, with only a few successful studies reported (Zhao et al., 2008; Beyene et al., 2016).

4.2.4.3 Transgenic Approaches

The rapid advancements in biotechnology and genome sequencing have enabled the development of diverse techniques for identifying and manipulating genes involved in key biological processes, including drought responses. Current research on engineering drought-tolerant plants primarily focuses on introducing one or more genes that play roles in signaling and regulatory pathways or encode enzymes involved in the synthesis of functional and structural protectants such as osmolytes, antioxidants, and stress tolerance-related proteins (Vinocur & Altman, 2005; Khan et al., 2019). Genetic engineering offers a faster and more efficient approach to improving drought tolerance compared to conventional breeding, significantly reducing the time required for trait development (Martignago et al., 2020). Using genetic engineering techniques, both natural and synthetic genes, as well as TFs, have been introduced into maize to enhance drought tolerance (Parmar et al., 2017).

Numerous drought-responsive genes and TFs have been identified, including:

- TsVP and BetA from *Thellungiella halophila* and *Escherichia coli*
- ZmPLC1, ZmVPP1, ZmTIP1, and ZmNAC111 from maize (Wang et al., 2008, 2016; Mao et al., 2015; Zhang et al., 2020a)
- SbER1-1 and SbER2-1 from sorghum (Li et al., 2019)
- LOS5 from *Arabidopsis thaliana* (Lu et al., 2013)
- TsCBF1 from *Thellungiella halophila* (Zhang et al., 2010)

Overexpression of these genes via the transgenic approach has been shown to improve drought tolerance in maize (Jia et al., 2020; Muppala et al., 2021; Yu et al., 2021). A major focus of genetic engineering for drought tolerance in maize has been on DREB proteins, a subfamily of the APETALA2/ET-responsive element binding factor (AP2/ERF) family, as well as mitogen-activated proteins (Bidhan et al., 2011). Various genes have been introduced into maize to enhance drought resistance:

- Shou et al. (2004) incorporated *Nicotiana* Protease Kinase (NPK1) via genetic transformation, which induces HSPs and glutathione-S-transferases (GSTs) to protect the photosynthetic machinery under drought stress.

- Quan et al. (2004) transformed an elite maize inbred line DH4866 with a key *E. coli* enzyme encoding choline dehydrogenase, which is linked to maize abiotic stress responses. Transgenic plants accumulated more glycine betaine (GB), leading to improved GY and drought resistance compared to non-transgenic plants.
- Amara et al. (2013) enhanced drought tolerance in maize by overexpressing the group LEA Rab28 gene, which resulted in higher Rab28 protein accumulation, stability, and improved water stress tolerance.
- He et al. (2018) reported that overexpression of ZmPYL3, ZmPYL9, ZmPYL10, and ZmPYL13 genes played a crucial role in enhancing ABA signaling, proline accumulation, and the expression of drought-related marker genes, leading to increased drought resistance in transgenic maize.
- Habben et al. (2014) used a gene-silencing approach to modulate ET biosynthesis in maize, which resulted in significantly higher GY and increased kernel number per ear under drought stress.
- Liu et al. (2015) cloned and analyzed the maize ZmSDD1 gene, revealing that its overexpression in transgenic maize resulted in a 6.68-fold increase in expression levels, reduced stomatal density, and an 80% higher survival rate compared to wild-type plants.

Monsanto, in collaboration with BASF, developed the first biotechnology-derived drought-tolerant maize variety, MON 87460, which expresses a bacterial cold shock protein B (CspB). This transgenic maize variety demonstrated a 6% increase in GY under drought conditions (Nemali et al., 2015). In addition, ZmNF-YB2, a homolog of AtNF-YB, was found to be constitutively expressed in transgenic maize and outperformed Arabidopsis NF-YBs in terms of drought survival and stress tolerance.

Field trials have demonstrated that transgenic maize plants exhibit enhanced drought resistance due to several physiological advantages, including higher stomatal conductance, cooler leaf temperatures, increased chlorophyll content, and delayed onset of senescence (Nelson et al., 2007). Recently, Muppala et al. (2021) introduced rpk and nced genes into elite maize inbred lines using LeaP and SalT promoters to develop stable transgenic lines. The transgenic plants exhibited higher accumulation of ABA and significantly improved survival rates under drought stress compared to control plants. Improving drought tolerance in maize requires a multifaceted approach, integrating breeding, biotechnological, and agronomic strategies. Research on transgenic maize highlights genetic engineering as a valuable tool for mitigating the risks associated with drought stress, offering promising solutions for sustainable crop production in water-limited environments.

4.2.4.4 Genome Editing

Recent advancements in genome editing technologies have enabled precise and rapid modification of DNA sequences, facilitating the development of drought-tolerant germplasm. Tools such as meganucleases, zinc-finger nucleases (ZFN), transcription activator-like effector nucleases (TALEN), and the clustered regularly interspaced short palindromic repeat (CRISPR)/CRISPR-associated protein 9 (Cas9) system have significantly improved targeted gene modification in plants (Li et al., 2013; Cermák et al., 2015; Khalil, 2020). Among these, CRISPR-Cas9 has revolutionized plant and animal research as a powerful and precise genome editing tool (Shan et al., 2013; Wang et al., 2019). This technology has been widely adopted in several crop species, including Arabidopsis (Fauser et al., 2014), rice (Hu et al., 2016), wheat (Zhang et al., 2016), potato (Andersson et al., 2018), tomato, and various fruit crops to enhance traits such as yield, disease resistance, and tolerance to abiotic stresses like drought and salinity (Wang et al., 2019).

CRISPR-Cas technology has been instrumental in identifying gene functions under various biotic and abiotic stresses by creating targeted gene knockouts or knockdowns. In maize, recent studies have explored its potential for enhancing drought tolerance. For instance, Shi et al. (2017) used CRISPR-Cas-mediated advanced breeding to generate novel variants of ARGOS8, a negative regulator of ET responses. The results demonstrated that ARGOS8 variants improved GY under drought stress compared to wild-type plants. This study highlights the potential of modifying a single native gene to maintain yield stability under drought conditions, while also allowing such plants to be classified as non-GM crops. However, drought tolerance is a complex trait involving multiple genes and metabolic pathways. Achieving practical improvements requires molecular techniques capable of simultaneous multi-gene editing. Advances in Golden Gate cloning and Gibson assembly have facilitated the multiplexing of multiple single-guide RNAs (sgRNAs) into a single CRISPR/Cas9 expression vector, enhancing genome editing efficiency (Silva & Patron, 2017). Another promising development is haploid-induction-mediated genome editing (IMGE), also known as HI-Edit technology. In this approach, a haploid inducer (HI) line also serves as the genome editor. When crossed with a target plant, genome edits occur in the zygote, followed by haploidization (uniparental genome elimination) (Malenica et al., 2021). This method accelerates elite breeding line development by avoiding the need for selfing or backcrossing to remove the transgene encoding the editing machinery. Despite its scientific promise, CRISPR-edited crops face socio-political challenges, including public perception and regulatory policies (Scheben & Edwards, 2017). Addressing these concerns will be crucial for the widespread adoption of genome editing in crop improvement.

4.3 LOW-TEMPERATURE STRESS TOLERANCE: IMPACTS AND IMPROVEMENT

Maize (*Zea mays* L.) is one of the world's most significant cereal crops, serving as a staple for food, a key economic commodity, and a primary source of livestock feed (Wang et al., 2020a). However, due to its tropical and subtropical origins, maize is highly sensitive to LTs and susceptible to LT stress (yyGreaves, 1996). While research has indicated that cold temperatures can impact grain filling (Chen & Tang, 2012), the most critical effects of LT stress occur during early developmental stages, including seed germination, seedling establishment, and overall vegetative growth. These early stage disruptions ultimately lead to reduced GY (Li et al., 2019b). In particular, LT stress damage during early spring is a major meteorological threat to maize production, especially in northern latitudes and high-altitude mountainous regions, where temperature fluctuations are more pronounced (Zhang, 2014). Cold stress significantly reduces both the seed germination rate and seedling vigor in maize (*Zea mays* L.) (Zhang et al., 2020a). During the critical water absorption phase (imbibition), exposure to LTs compromises cell membrane permeability, leading to the leakage of essential cellular components (Hussain et al., 2018). Additionally, LT stress can disrupt the ultrastructure of embryonic root meristem cells, hindering root development and overall seedling establishment (Farooq et al., 2009) (Figure 4.2).

Maize seedling growth slows significantly when temperatures drop below 10°C and ceases entirely between 6°C and 8°C (Peng et al., 2016). Prolonged exposure to lower temperatures can cause irreversible damage to seedling cells and tissues. LT stress during early growth stages adversely affects several physiological traits, including plant height, root elongation, nutrient uptake efficiency, chlorophyll content, and the net photosynthetic rate. These impairments ultimately lead to growth inhibition, leaf chlorosis, wilting, necrosis, and in severe cases, seedling mortality (Chen & Tang, 2012; Yang et al., 2018). Beyond these direct effects, LT stress can also weaken maize seedlings indirectly by increasing their susceptibility to infections from soil-borne pathogens, further compromising plant health and vigor (Juurakko et al., 2021). The combined direct and indirect consequences of LT stress can significantly reduce both yield and grain quality. Enhancing cold tolerance in

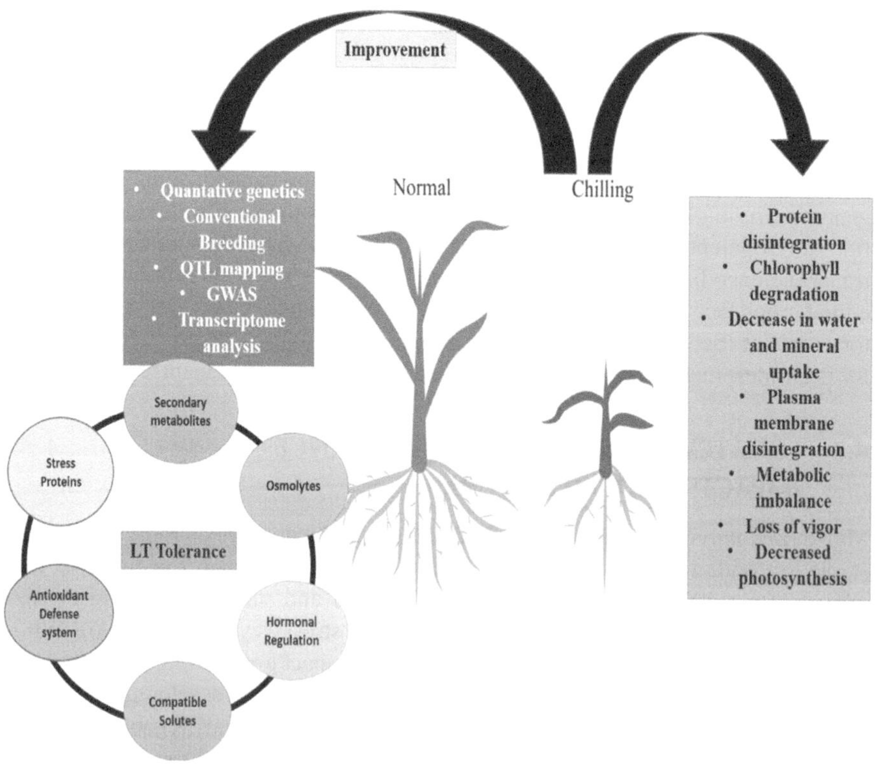

FIGURE 4.2 Low-temperature stress tolerance and its improvement in maize.

maize is critical for enabling early season sowing, allowing plants to endure unexpected cold snaps in early spring. Early planting can extend the vegetative growth period, facilitating greater biomass accumulation and improving overall productivity (Aydinoglu, 2020). In this section of the chapter, we summarize recent advancements in the molecular, physiological, genetic, and genomic mechanisms underlying cold tolerance in maize, providing a theoretical foundation for molecular breeding strategies aimed at improving cold resilience in this vital crop (Table 4.2).

TABLE 4.2
Physiological and Molecular Responses of Maize to Low-Temperature (LT) Stress

Category	Response/Component	Mechanism/Effect under LT Stress	References
Growth & Development	Seed Germination, Seedling Vigor	Inhibited germination, poor seedling establishment, damaged root meristems, reduced plant height, and chlorosis under <10°C	Li et al., 2019b; Peng et al., 2016; Farooq et al., 2009
Osmoregulation	Soluble sugars, proteins, proline	Accumulation helps lower water potential, protect cellular structures, and maintain metabolic stability	Zhang, 2014; Guan et al., 2009; Ma et al., 2015a
Antioxidant Enzymes	SOD, CAT, POD, APX	Scavenge ROS to prevent oxidative damage; higher activity in cold-tolerant lines	Nadarajah, 2020; Peng et al., 2016; Wei et al., 2014
	MDA & Relative Conductivity	Indicators of membrane damage; lower in cold-tolerant varieties	Zhang, 2014; Wang et al., 2006
Hormonal Regulation	ABA, GA, IAA, SA, BL	ABA increases during cold stress; exogenous application of ABA, $SA+H_2O_2$, and BL enhances cold tolerance	Zhang, 2011; Li et al., 2017; Sun et al., 2020
Mineral Nutrition	Mn, Zn, P, Si	Enhances nutrient uptake (especially P); supports ROS detoxification and hormonal balance	Gómez-Muñoz et al., 2018; Moradtalab et al., 2018
Photosynthesis	Chlorophyll content, PSII, G6PD, sugar ratio	Reduced chloroplast function and enzyme activity; cold-tolerant lines maintain higher photosynthetic efficiency	Fracheboud et al., 2002; Duran Garzon et al., 2020
Molecular Pathways	ICE-CBF-COR, ZmmICE1	Conserved cold tolerance pathway; ZmDREB1A and related genes upregulated under cold	Wang & Dong, 2009; Lu et al., 2017
Transcription Factors	ZmMYB31, ZmMYB-IF35, ZmMYB53	Positively regulate cold-inducible genes (e.g., CBFs, CORs); ZmMYB53 is cold-specific	Chen et al., 2018; Meng & Sui, 2019
MAPK Cascade	ZmMPK5, ZmMKK4, ZmMPK17, ZmMKK1, ZmMPK8-RR1	MAPK genes regulate antioxidant defenses and cold-responsive gene expression	Kong et al., 2011; Pan et al., 2012; Zeng et al., 2021

4.3.1 Physiological Impact

LT stress triggers a cascade of physiological responses in maize, including osmotic adjustments, ROS accumulation, hormonal imbalances, impaired mineral nutrient uptake, and reduced photosynthetic efficiency. To withstand these adverse conditions, plants must preserve cellular function and structural integrity by stabilizing cell membranes and protecting biologically active proteins, thereby sustaining essential physiological processes (Ritonga & Chen, 2020). Although maize has evolved acclimation mechanisms to mitigate cold stress, several agronomic and molecular strategies have been developed to further enhance cold tolerance and minimize its detrimental effects on seedling growth and development.

4.3.1.1 Osmoregulatory Substances

Osmotic adjustment is a crucial physiological mechanism that enables plants to withstand various environmental stresses. Under adverse conditions, osmotic regulators help maintain cell turgor and enhance the ability of tissues to retain water, with their accumulation being positively correlated with plant stress resistance (Farhangi-Abriz & Torabian, 2017). By lowering cellular water potential, these osmolytes prevent excessive water loss while simultaneously stabilizing macromolecular structures and preserving cellular integrity (Li et al., 2013). The primary osmotic adjustment compounds in plants include soluble sugars, soluble proteins, and free proline, all of which play essential roles in stress adaptation (Zhang, 2014). Soluble sugars, due to their strong hydrophilic properties, play a critical role in mitigating stress-induced cellular damage by lowering water potential and preserving the activity of essential biological macromolecules (Guan et al., 2009). Similarly, soluble proteins contribute to stress tolerance by scavenging ROS and stabilizing cell membrane integrity (Nadarajah, 2020). Proline, a hydrophilic amino acid, serves as a key osmoprotectant, maintaining metabolic stability in the cytoplasm under stress conditions (Guan et al., 2009). LT stress has been shown to induce the expression of *ZmP5CS1*, a key gene involved in proline biosynthesis, particularly in immature maize embryos (Ma et al., 2015a). Several studies have demonstrated that exposure to LTs leads to increased accumulation of soluble sugars, proline, and soluble proteins in maize seedlings. Notably, LT-tolerant maize varieties exhibit a greater relative increase in these osmotic adjustment substances compared to more sensitive varieties, with the magnitude of difference becoming more pronounced at lower temperatures (Li & Fang, 2018; Zhao et al., 2020).

In summary, LT-tolerant maize maintains cellular water absorption capacity and mitigates LT-induced damage by accumulating higher levels of osmotic adjustment compounds in response to stress. This enhanced osmotic regulation plays a pivotal role in improving maize resilience under cold conditions.

4.3.1.2 Antioxidant Enzyme System

Cold stress disrupts cellular homeostasis in plants, leading to excessive accumulation of ROS. High levels of ROS can be detrimental to cellular integrity, primarily through lipid peroxidation, which damages the cell membrane system (Nadarajah, 2020). To counteract ROS-induced stress, plant cells activate a complex antioxidant defense system. This protective mechanism involves the coordinated action of enzymatic antioxidants such as superoxide dismutase (SOD), catalase (CAT), peroxidase (POD), and ascorbate peroxidase (APX), which work synergistically to scavenge ROS and mitigate oxidative damage (Nadarajah, 2020). By maintaining redox balance, maize cells can stabilize membrane structures, protect cellular components, and preserve metabolic functions, thereby enhancing their resilience to cold stress. Numerous studies have demonstrated that cold stress enhances the activity of key antioxidant enzymes, including SOD, POD, and CAT, in maize. The enzymatic activity levels and the relative expression of genes encoding these antioxidants are positively correlated with the cold tolerance of maize inbred lines (Wei et al., 2014; Yang et al., 2016; Li & Fang, 2018).

For instance, early research found that LTs induce the expression of the *APX* gene, with cold-tolerant maize exhibiting higher APX activity than cold-sensitive varieties (Pinhero et al., 1997). Similarly, another study reported that while cold stress led to a decline in protective enzyme activity in both cold-tolerant and cold-sensitive maize inbred lines, the reduction was significantly less pronounced in the cold-tolerant lines (Peng et al., 2016). These variations in antioxidant enzyme responses may be attributed to multiple physiological and genetic factors. However, it is evident that cold-resistant maize lines generally possess a more robust antioxidant defense system compared to cold-sensitive lines, contributing to their enhanced ability to withstand oxidative damage under LT stress. Malondialdehyde (MDA) is a key byproduct of polyunsaturated lipid peroxidation in cell membranes and serves as a biochemical marker for oxidative stress. As a highly reactive compound, MDA can inhibit the activity of protective enzymes and exacerbate membrane lipid peroxidation, leading to cellular damage. When the integrity of the cell membrane is compromised, the leakage of intracellular electrolytes increases, resulting in a surge in relative conductivity (Wang et al., 2006). Consequently, MDA levels and relative conductivity are negatively correlated with cold tolerance in maize, with higher values indicating greater membrane damage and increased susceptibility to cold stress (Zhang, 2014). Studies have shown that prolonged exposure to LTs leads to a progressive increase in both MDA accumulation and relative conductivity. However, cold-tolerant maize varieties exhibit less membrane damage, with only a moderate rise in MDA content and electrolyte leakage compared to cold-sensitive lines (Zhang, 2014; Peng et al., 2016).

4.3.1.3 Plant Hormones

Plants meticulously regulate hormone levels to adapt to environmental changes (Lamers et al., 2020). ABA, a key abiotic stress hormone, enhances plant resilience by mediating responses to various stresses, including cold (Qin et al., 2021). Notably, cold-tolerant maize lines exhibit significantly higher ABA levels in their roots compared to cold-sensitive lines. Additionally, exogenous ABA application has been shown to enhance maize seed germination under LTs (Zhang, 2011). Gibberellic acid (GA) also contributes to cold resistance in plants, although its role is less pronounced than that of ABA (Eremina et al., 2016; Rihan et al., 2017). Research indicates that cold stress leads to a reduction in gibberellin (GA) and indole-3-acetic acid (IAA; auxin) levels in maize seedling roots, while ABA levels progressively increase in response to the stress (Janowiak et al., 2002; Zhang, 2011; Wei et al., 2014). SA is another key plant hormone involved in cold tolerance across various species, including maize. Its primary physiological role in cold stress adaptation is regulating antioxidant enzyme activity (Farooq et al., 2009). Under LTs, an increase in endogenous SA biosynthesis is closely associated with enhanced antioxidant enzyme activity in maize seeds and seedlings during early growth stages (Wang et al., 2013). Cold stress in maize seedlings was not alleviated by SA seed treatment alone (Gómez-Muñoz et al., 2018). However, a combined application of SA and H_2O_2 at LTs shortened germination time, enhanced seedling vigor, reduced cold-induced damage, and increased both antioxidant enzyme activity and the expression of related genes, ultimately improving cold tolerance (Li et al., 2017). Similarly, exogenous application of brassinolide (BL) enhanced germination rates, mitigated cold damage, and boosted antioxidant enzyme activity, leading to increased plant biomass (Sun et al., 2020). These findings suggest that manipulating plant hormone levels is an effective strategy to mitigate cold stress damage and support seedling growth during later developmental stages.

4.3.1.4 Mineral Nutrients

Cold tolerance in plants is strongly associated with mineral nutrient levels (Waraich et al., 2012). In maize, cold stress-induced growth inhibition is at least partially due to reduced nutrient uptake from the soil (Gómez-Muñoz et al., 2018). For instance, the absorption of potassium (K) and orthophosphate (Pi) by maize roots is particularly sensitive to low soil temperatures (Bravo et al., 1981). Seed treatments with manganese (Mn) and zinc (Zn) have been shown to mitigate the adverse effects of cold stress, leading to increased biomass production in high-phosphorus (P) soils but not in low-P soils. This suggests that Mn/Zn treatment enhances phosphorus uptake under cold conditions (Gómez-Muñoz et al., 2018). Additionally, Mn and Zn serve as essential cofactors for enzymes involved in ROS detoxification during cold stress. Silicon (Si) application is another effective strategy for improving cold tolerance in maize seedlings during early growth. Si seed treatments help restore hormonal balance disrupted by cold stress and maintain micronutrient homeostasis (Moradtalab et al., 2018). Thus, optimizing plant nutrition is a valuable approach for alleviating the detrimental effects of cold stress and supporting healthy seedling development.

4.3.1.5 Photosynthesis

Cold stress significantly reduces photosynthesis in maize (Fracheboud et al., 2002). This decline can be attributed to various factors, including impaired chloroplast development, alterations in pigment composition, damage to PSII reaction centers, and reduced activity of carbon cycle enzymes. The extent of photosynthetic reduction under LTs is also an indicator of maize cold tolerance (Hund et al., 2008). Research by Fracheboud et al. (2002) found that cold-tolerant maize lines exhibit a higher chlorophyll a/b ratio than cold-sensitive lines. Additionally, Duran Garzon et al. (2020) reported that cold tolerance is associated with increased chlorophyll content, elevated glucose-6-phosphate dehydrogenase (G6PD) activity, and a higher sucrose-to-starch ratio. Given these correlations, photosynthesis-related traits have been widely used as indicators to assess cold tolerance in maize.

4.3.2 Impact on Molecular Network

The ICE-CBF-COR pathway has been studied in maize to understand its role in cold tolerance. However, the functional validation of most maize cold-responsive genes has been conducted through expression in heterologous species, such as Arabidopsis or tobacco, rather than in maize itself. Several maize CBF TF genes, including *ZmDREB1A*, *ZmDREB2A*, *ZmDBP3*, and *ZmDBP4*, are induced by cold stress. Overexpression of *ZmDREB1A*, *ZmDBP3*, and *ZmDBP4* in Arabidopsis led to the upregulation of cold-inducible genes, resulting in enhanced cold tolerance (Qin et al., 2004; Wang & Dong, 2009; Wang et al., 2011). In contrast, *ZmDREB2A* overexpression improved thermotolerance but did not confer cold tolerance (Qin et al., 2007). An LT-associated gene, *ZmmICE1*, was isolated from *Zea mays ssp. mexicana L.*, a close relative of maize. Ectopic expression of *ZmmICE1* in the Arabidopsis *ice1-2* mutant enhanced cold tolerance (Lu et al., 2017). Additionally, transformation of the forage maize line SAUMZ1 with the Arabidopsis *AtCBF1* gene reduced relative electrolyte leakage compared to wild-type plants, improving cold tolerance (Xiang et al., 2012). These studies suggest that the ICE-CBF-COR pathway is conserved across plant species. Beyond the ICE and CBF genes, numerous other regulators, including TFs and protein kinases, have been functionally linked to cold tolerance (Weckwerth et al., 2015; Erpen et al., 2018; Kimotho et al., 2019).

MYB TFs play a crucial role in plant responses to various stresses, including cold stress (Wu et al., 2019). Unlike *AtMYB15* in Arabidopsis, which acts as a negative regulator of *CBF* gene transcription, two characterized maize MYB TFs function as positive regulators of *CBF* genes. The *ZmMYB31* gene, encoding an R2R3-MYB TF, is upregulated in response to LTs. Additionally, *ZmMYB-IF35*, another nucleus-localized R2R3-MYB TF, was identified by Meng and Sui (2019). Its expression is induced by cold stress in the cold-tolerant maize inbred line M54. *ZmMYB-IF35* positively regulates the expression of stress-responsive genes such as *AtCBF2*, *AtCBF3*, *AtCOR1*, and *AtCOR2*, contributing to enhanced cold tolerance. A broader analysis of 46 *ZmMYB* genes under different abiotic stresses revealed that 22 of them respond to various stress treatments. Among these, six *ZmMYB* genes were responsive to cold stress, but only *ZmMYB53* exhibited exclusive induction by cold stress (Chen et al., 2018).

Mitogen-activated protein kinases (MAPKs or MPKs) are a family of serine/threonine protein kinases involved in various biological processes, including stress signal transduction and plant development (Kong et al., 2013). Several MAPK-related genes have been implicated in maize cold tolerance. In maize leaves, *ZmMPK5* plays a role in plant recovery from cold stress (Berberich et al., 1999). Kong et al. (2011) identified *ZmMKK4*, a group C MAPKK gene, from the root system of the maize variety "Zhengdan 958." Its expression was upregulated under LTs, and overexpression of *ZmMKK4* in Arabidopsis enhanced cold tolerance, indicating its role as a positive regulator of cold stress responses. Similarly, Pan et al. (2012) identified *ZmMPK17*, a group D MAPK gene induced by cold stress. Overexpression of *ZmMPK17* in tobacco (*Nicotiana tabacum*) enhanced cold tolerance by modulating the antioxidant defense system. Additionally, Cai et al. (2014) characterized *ZmMKK1*, a group A

MAPKK gene from "Zhengdan 958," whose ectopic expression in tobacco improved cold tolerance, further supporting its role in cold stress response. More recently, Zeng et al. (2021) identified a novel cold regulation pathway in maize, *ZmMPK8-ZmRR1-ZmDREB1.10/ZmCesA2*. Overexpression of *ZmRR1* enhanced cold tolerance by upregulating the expression of its downstream genes, *ZmDREB1.10* and *ZmCesA2* (Cellulose Synthase 2), suggesting that *ZmRR1* functions as a positive regulator of cold tolerance. Currently, this pathway represents the most well-characterized molecular mechanism underlying cold tolerance in maize, providing critical insights into maize stress adaptation strategies.

4.3.3 Approaches for Improvement of LT Stress Tolerance in Maize

Cold tolerance in maize is a complex quantitative agronomic trait regulated by multiple genes (Turk et al., 2019). In recent years, various genetic and genomic approaches have been employed to dissect the molecular and genetic components underlying this trait. These studies have provided valuable insights into the key regulatory mechanisms governing cold tolerance, aiding in the development of more resilient maize varieties.

4.3.3.1 Quantitative Genetics

Quantitative genetic analyses have revealed that cold tolerance in maize is influenced by multiple genetic and environmental factors, including heterosis, general and specific combining abilities, reciprocal maternal and non-maternal effects, additive effects, genotype, growth stage, and environmental conditions.

For instance, Hodges et al. (1997a, b) conducted studies on maize cold tolerance at both germination and seedling stages using twelve hybrids derived from a complete diallel cross of four inbred lines. Their findings demonstrated significant variation in seed germination rates, CAT and APX activities in leaves at the 3-leaf stage, as well as general and specific combining abilities for dry matter accumulation in leaves at the 4-leaf stage. Similarly, Revilla et al. (2000) proposed that cold tolerance is governed by additive, dominant, and maternal genetic effects. Additionally, Yan et al. (2017) found that maize exhibits significant heterosis for cold tolerance at the seedling stage, further emphasizing the complex genetic architecture of this trait. Neta et al. (2020) conducted partial diallel cross experiments using three cold-tolerant and three cold-sensitive maize lines to examine heterosis, general and specific combining abilities, reciprocal maternal and non-maternal effects, as well as the expression of key antioxidant enzymes, including CAT, APX, and SOD. Their findings revealed heterosis and reciprocal effects on germination under cold stress, with non-additive genetic effects playing a more significant role in cold tolerance.

The genetic basis of cold tolerance varies depending on the specific plant materials and traits studied (Frascaroli & Revilla, 2018). Each maize growth stage—germination, emergence, and early seedling development—is governed by an independent genetic model, suggesting that cold tolerance may also be regulated separately at different developmental stages (Frascaroli & Revilla, 2018). Adding to this complexity, some studies have reported significant gene-environment interactions influencing the expression of cold tolerance in maize (Presterl et al., 2007), highlighting the multifaceted nature of this trait.

> Findings from these studies contribute to a deeper understanding of the genetic mechanisms underlying maize cold tolerance at the molecular level. This knowledge can aid in the development of breeding strategies aimed at enhancing cold resilience in maize, ultimately improving crop performance in low-temperature environments.

4.3.3.2 QTL Mapping

QTL mapping serves as a powerful approach for identifying and manipulating genetic loci associated with complex and agronomically significant traits in crops. Recent studies have employed various mapping populations, including F2:3 families, recombinant inbred lines (RILs), and backcross (BC) populations, to investigate QTLs governing cold tolerance in maize.

These populations are typically derived from crosses between cold-sensitive and cold-tolerant inbred lines to maximize phenotypic variation in response to cold stress. To assess cold tolerance, researchers have evaluated multiple physiological traits, such as seed germination rate, root morphology, and seedling photosynthetic efficiency, under both normal and cold stress conditions (Table 4.2). These studies have identified numerous molecular marker loci linked to QTLs or genes that regulate cold tolerance, offering valuable resources for breeding cold-tolerant inbred lines and hybrids. As summarized in Table 4.2, the genomic loci associated with cold tolerance are distributed across nearly all maize chromosomes. However, the majority of these QTLs have yet to be fine-mapped or functionally characterized as cold tolerance genes, limiting deeper insights into the underlying molecular mechanisms. A primary challenge in this regard is the extensive effort required for gene mapping in large populations.

In recent years, joint analysis approaches utilizing advanced populations and genomic tools have gained popularity for refining candidate gene identification. Notably, this strategy was successfully applied to pinpoint the major cold-regulating gene *Zm00001d002729* (Jin et al., 2021), demonstrating its potential for broader application in future studies to enhance our understanding of cold tolerance in maize.

4.3.3.3 GWAS

In addition to biparental segregating populations, natural populations comprising several hundred maize inbred lines have been utilized for genetic analyses of LT tolerance through GWAS (Frascaroli & Revilla, 2018). While GWAS examines many of the same cold-responsive physiological traits as QTL mapping—such as leaf color, root length, and seed germination-related traits (Tables 4.1 and 4.2)—it provides higher mapping resolution and accuracy due to the greater genetic diversity present in the population (Lipka et al., 2015). The first GWAS for dissecting cold tolerance traits in maize was conducted by Strigens et al. (2013). Using a diverse germplasm collection of 375 inbred genotypes genotyped with 56K single-nucleotide polymorphisms (SNPs), this study identified 19 significant association signals. These loci collectively accounted for 5.7% to 52.5% of the phenotypic variance in cold-related traits, including early growth performance and chlorophyll fluorescence, highlighting the potential of GWAS in uncovering key genetic factors underlying cold tolerance. Huang et al. (2013) conducted a GWAS using 125 maize inbred lines to investigate ten traits associated with cold tolerance during the germination and seedling stages. This study identified 43 SNPs linked to cold tolerance, with 31 of these SNPs corresponding to 40 predicted candidate genes. Similarly, Yan et al. (2017) developed a test-cross association mapping population by crossing the cytoplasmic male sterile parental line S-Mo17 with 338 diverse inbred lines. GWAS on these test crosses revealed 19 significant SNPs associated with cold tolerance-related traits.

Further advancing GWAS applications in maize, Hu et al. (2017) analyzed seed germination traits under both normal and LT conditions in a panel of 282 inbred lines. This study identified 17 SNPs associated with cold tolerance, seven of which were located within candidate genes. Expanding on these findings, Zhang et al. (2020a) examined a population of 222 maize inbred lines and identified 30 SNPs linked to cold tolerance during seed germination. Fourteen of these SNPs were directly associated with candidate genes, which were subsequently validated through gene expression analysis, reinforcing their potential role in maize cold tolerance.

Zhang et al. (2021) investigated germination-related traits in 300 maize inbred lines under LT conditions, identifying 15 significant SNPs through GWAS analysis. Notably, three genomic loci were consistently associated with multiple traits, suggesting their potential role in cold tolerance. Similarly, Yi et al. (2021) conducted a large-scale GWAS using 836 maize inbred lines, uncovering 187 significant SNPs that mapped to 159 genomic regions regulating seed emergence and early growth traits. Despite its advantages in achieving higher genetic mapping resolution, GWAS are not without limitations. A major concern is that many cold-associated SNP markers identified in these studies are located in non-coding regions, which are generally believed to regulate gene expression. However, the specific genes they influence remain largely unknown. Additionally, the presence of large structural variations across different maize inbred genomes complicates genetic mapping. The B73 reference genome, commonly used for SNP annotation, may not fully capture the genetic diversity of other inbred lines. As a result, some unidentified genes may be located near or even within cold-associated SNP regions, potentially leading to misinterpretation of results and incomplete identification of causal genes.

4.3.3.4 Transcriptomic Analysis

Transcriptome analyses have become a valuable tool for investigating the molecular responses of maize to cold stress and identifying candidate genes associated with cold tolerance. Li et al. (2016) utilized RNA-seq to compare the transcriptomes of a freezing-tolerant maize line (KR701) and a freezing-sensitive line (Hei8834) before and after cold treatment at the seedling stage. This analysis identified 948 differentially expressed genes (DEGs), with Gene Ontology (GO) enrichment revealing an overrepresentation of terms related to "binding functions," "protein kinase activity," and "peptidase activity," suggesting their involvement in cold stress responses.

Building on this, Li et al. (2019b) conducted an RNA-seq analysis on cold-tolerant (M54) and cold-sensitive (753F) inbred lines subjected to cold stress at the seedling stage. The results showed a greater number of DEGs in M54 compared to 753F at both 4 and 24 hours of cold exposure, indicating that the cold-tolerant line exhibited a more active cold-responsive signaling network.

Further expanding on transcriptomic insights, Li et al. (2020) analyzed the response of maize B73 seedlings to different LT conditions. This study identified 5,358, 5,485, and 5,312 DEGs under cold stresses of 4°C, 10°C, and 16°C, respectively. Among these, the expression of five key genes, including *ZmDERB1*, was significantly upregulated, highlighting their potential roles in cold stress adaptation. Frey et al. (2020) investigated the transcriptomic responses to cold stress in 21 double haploid (DH) maize lines selected from the flint landrace *Petkuser Ferdinand rot* based on their cold tolerance. Among these, 11 lines were classified as cold-resistant and 10 as cold-sensitive. RNA-seq analysis conducted after both control and cold treatments identified 148 DEGs associated with cold exposure, 3,254 DEGs linked to cold tolerance, and 563 DEGs related to growth rate under LT conditions. Zhang et al. (2020a) further utilized RNA-seq to validate the correlation between candidate genes and LT tolerance. Their study identified 10 DEGs within the linkage disequilibrium regions of a GWAS analysis, two of which were implicated in cold signal transduction and the regulation of cell membrane fluidity—key mechanisms for cold adaptation. Expanding on the role of transcriptomics in cold tolerance, Li et al. (2021) analyzed gene expression changes in seeds of three sweet corn near-isogenic lines (NILs) and their parental lines under cold stress. This study identified 20 DEGs strongly associated with LT germination, with a gene encoding UDP-glucosyltransferase hypothesized to play a critical role in cold germination in sweet corn.

Waititu et al. (2021) conducted a comparative transcriptomic analysis of maize seedlings from 24 cold-tolerant and 22 cold-sensitive inbred lines subjected to cold stress. Their study uncovered 2,237 DEGs, including 147 TFs belonging to 32 families such as *MYB, ERF, NAC, WRKY, bHLH, MIKC MADS*, and *C2H2*, highlighting the extensive regulatory networks involved in maize cold tolerance. Yu et al. (2021) examined the leaf transcriptomic responses of two maize inbred lines with contrasting cold tolerance under a time-series cold treatment. Their findings revealed that the cold-tolerant line B144 mitigates cold stress by actively regulating stomatal opening and protecting photosystem II from photooxidation through the upregulation of *D1* protein-related genes. In contrast, the cold-sensitive line Q319 failed to close its stomata in response to cold, making it more susceptible to stress-induced damage.

Beyond RNA-seq analyses, other genomic tools have been employed to investigate transcriptomic changes in maize under cold stress. Rodríguez et al. (2013) utilized microarray hybridization to analyze the chlorophyll-deficient and chlorophyll-containing sections of leaves from the maize inbred line A661, which exhibits a cold-induced albino phenotype. This study identified 1,002 differentially expressed transcripts between the two tissue types, with DEGs classified into 23 functional categories, including genes involved in the tetrapyrrole biosynthesis pathway and photosynthesis. Similarly, Di Fenza et al. (2017) performed microarray analysis on four maize varieties—two cold-tolerant and two less cold-tolerant—to identify genes differentially expressed under chilling conditions. These findings further highlight the diverse molecular mechanisms regulating maize cold tolerance and underscore the potential of transcriptomic studies in uncovering key genetic determinants of stress adaptation. Di Fenza et al. (2017) identified a total of 64 DEGs in the two chilling-tolerant maize varieties, whereas no significant gene expression changes were observed in the less cold-tolerant lines, suggesting a stronger transcriptional response in tolerant genotypes. In another study, Yang et al. (2011) used cDNA-AFLP analysis to investigate gene expression changes in maize under cold stress. Their study identified three candidate genes—*ZmMAPKKK, ZmCLCD,* and *ZmRLK*—that may play crucial roles in cold response pathways. To explore the regulatory role of microRNAs (miRNAs) in maize cold tolerance, Aydinoglu (2020) performed a miRNA microarray analysis (miRNome profiling) on seedlings of the maize hybrid ADA313 subjected to cold stress. The results revealed that 24, 6, and 20 miRNAs were differentially regulated in the meristem, elongation zone, and mature zone, respectively. These findings emphasized the significance of miRNA-mediated gene regulation in maize's response to cold stress. Expanding on transcriptomic research, Zhou et al. (2022) combined large-scale bioinformatics analysis with RNA-seq to examine transcriptomic changes in maize seedlings of three inbred lines (B73, Mo17, and W22) and their corresponding F1 hybrids (B73×Mo17, W22×B73, and W22×Mo17) under cold and heat stress conditions. Their study identified numerous stress-responsive DEGs across different maize genotypes and leveraged F1 hybrid data to distinguish between cis- and trans-regulatory mechanisms. By analyzing sequence variations in cis-elements, the study provided insights into how genetic differences influence gene responsiveness to stress. Moreover, the implementation of advanced computational modeling using convolutional neural networks (CNNs) offered a novel approach for predicting plant stress responses, advancing the potential for precision breeding of stress-resilient maize varieties. Transcriptomic analyses of maize cold stress responses have identified anywhere from tens to thousands of DEGs in individual studies. This striking variation suggests that the vast majority of DEGs do not overlap across studies, making it extremely challenging to pinpoint conserved genes with consistent roles in cold tolerance. Sowiński et al. (2020) systematically reviewed 13 independent studies examining transcriptomic changes in maize seedlings at the V2–V5 developmental stages under cold stress. Their meta-analysis revealed that fewer than 500 DEGs were consistently reported in more than one study, while the majority were unique to specific experiments. A key finding was that the nature of the cold stress applied influenced the transcriptomic response: four of the studies subjected maize plants to moderate LT stress, whereas the remaining nine used severe cold stress. The comparison of DEGs across these studies demonstrated that the gene expression changes induced by moderate and severe cold stress were fundamentally different.

These variations in gene expression across studies can be attributed to multiple factors, including genetic differences among maize genotypes, tissue-specific responses, developmental stage at sampling, experimental design, and the intensity of cold exposure. Despite these discrepancies, identifying "common genes" that consistently appear across multiple independent studies—after carefully controlling for confounding variables—offers a promising strategy for selecting strong candidate genes for further functional validation and breeding applications.

4.4 SALINITY STRESS: EFFECTS AND MANAGEMENT STRATEGIES

Maize (*Zea mays* L.) production is significantly constrained by both biotic and abiotic stressors, with abiotic stresses posing the most severe global threat to crop productivity. Among these, soil salinity stands out as a major limiting factor, adversely affecting maize growth and yield across all developmental stages—from germination through to seedling establishment, vegetative growth, and reproductive development. Salinity stress disrupts normal plant physiology by inducing osmotic stress, ion toxicity (particularly from sodium and chloride ions), and nutrient imbalances, which collectively impair cellular function and metabolic stability (Table 4.3).

TABLE 4.3
Effects of Salt Stress on Maize Growth, Physiology, and Yield Components

Developmental Stage	Aspect Affected	Salt Stress Effects	References
Seed Germination and Seedling Growth	Germination rate, vigor, MGT, plumule/shoot/root length	Delayed germination, reduced vigor index, inhibited root and shoot elongation (up to 78–95%); Na^+ and Cl^- toxicity impairs metabolism, osmotic balance, and water uptake	Aliu et al., 2015; Hassan et al., 2018; Khān & Weber, 2008; Khodarahmpour et al., 2012
	Ion imbalance	Accumulation of Na^+/Cl^- disrupts ion homeostasis and enzymatic activities	Murillo-Amador et al., 2002; Rasheed, 2009
	Water relations	Hyperosmotic soil conditions reduce seed imbibition and water uptake	Khān & Weber, 2008; Allen, 1994
	Seed size response	Small seeds show higher germination index under salt stress	Kaya et al., 2008
	Mitigation	Seed priming (e.g., with NaCl) improves germination under saline conditions	Farooq et al., 2006
Vegetative Growth	Shoot and root growth	Inhibited leaf initiation, expansion, and internode elongation; reduced leaf/stem biomass and dry weights; premature senescence	Akram et al., 2010; Ahmed et al., 2023; Ashrafuzzaman et al., 2002
	Cell expansion and division	Apoplastic acidification and Ca^{2+} displacement inhibit cell wall loosening and expansion	Hager, 2003
	Photosynthetic machinery	Reduced chlorophyll, carotenoids, Fv/Fm ratio; photoinhibition and pigment degradation	Deng et al., 2010;
	Enzyme activity	Inhibition of nitrate reductase and other enzymes in nitrogen assimilation pathways	Munns et al., 2006; Chattha et al., 2023
	Ion homeostasis	Na^+ and Cl^- accumulation in shoots; reduced K^+ transport; disrupted membrane selectivity	Kaya et al., 2010; Niu et al., 2012
	Carbon assimilation	Decreased CO_2 fixation due to impaired enzyme function and stomatal conductance	Wang et al., 2026

(Continued)

TABLE 4.3 *(Continued)*
Effects of Salt Stress on Maize Growth, Physiology, and Yield Components

Developmental Stage	Aspect Affected	Salt Stress Effects	References
Reproductive Stage	Grain yield, grain number and weight	Yield reduction (22–34%) due to impaired assimilate production and translocation; poor grain filling	Schubert et al., 2009; Hu et al., 2022
	Assimilate partitioning	Impaired flow from source (leaves) to sink (grains), leading to incomplete grain filling	Lohaus et al., 2000; Hiyane et al., 2010
	Stress timing impact	Yield losses vary with developmental stage at stress exposure and genotype sensitivity	Hasan et al., 2019

The extent of yield loss is influenced by the severity and duration of salt exposure, the developmental stage of the crop, soil properties, and the genetic makeup of the cultivar.

To cope with salinity, maize plants engage a range of intricate physiological, biochemical, and metabolic adjustments. These include modulation of stomatal conductance, photosynthetic efficiency, respiratory activity, transpiration rates, hormonal signaling, enzyme activity, and the synthesis of protective metabolites (Figure 4.3).

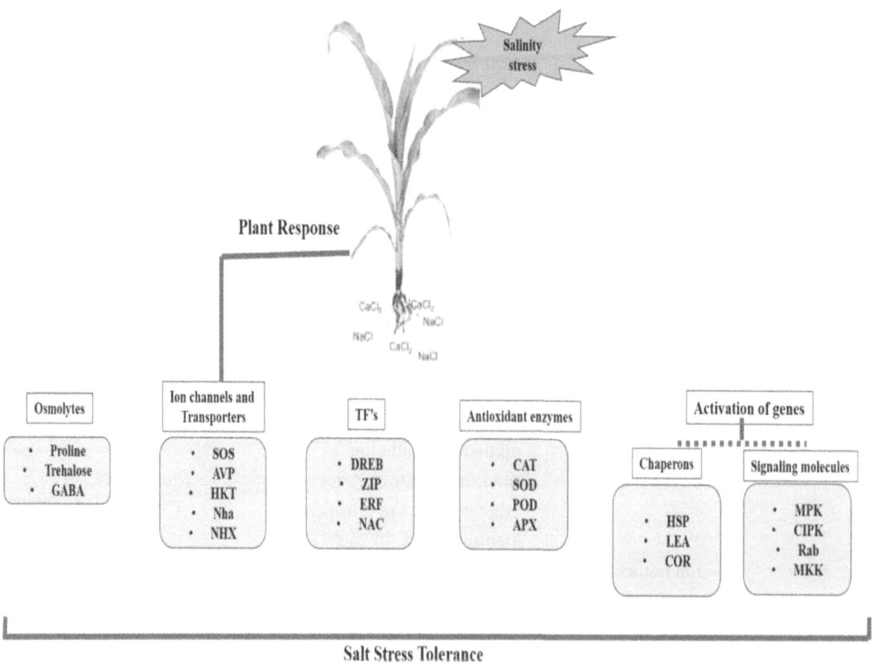

FIGURE 4.3 Salt stress tolerance in maize.

Understanding these adaptive responses provides critical insight into the mechanisms of salt tolerance. Accordingly, comprehensive studies on salt stress responses and tolerance mechanisms in maize are essential for the development of integrated and sustainable management strategies tailored to saline agricultural environments This section will therefore explore the multifaceted impacts of salt stress on maize at various growth stages, elucidate the underlying tolerance mechanisms, and outline agronomic practices aimed at enhancing productivity in salt-affected regions.

4.4.1 Effects of Salt Stress

4.4.1.1 Seed Germination and Seedling Growth

Salinity stress is a major abiotic constraint that adversely affects seed germination and early seedling development in maize. Elevated salt concentrations in the soil delay germination onset and suppress seedling vigor, often through complex interactions with environmental variables such as temperature and light (Bojović et al., 2010; Aliu et al., 2015). High salinity not only hinders seed germination and root elongation but also restricts water and nutrient uptake, thereby impeding vegetative growth (Grewal et al., 2004; Gong et al., 2018). The accumulation of toxic ions—particularly sodium (Na^+) and chloride (Cl^-)—within plant tissues disrupts ion homeostasis and interferes with fundamental physiological processes.

Under saline conditions, seeds encounter hyperosmotic stress, which reduces water uptake due to lowered soil osmotic potential, ultimately impairing imbibition—the initial phase of germination (Khān & Weber, 2008). Moreover, ion toxicity alters enzymatic activities essential for nucleic acid and protein metabolism, further inhibiting germination (Rasheed et al., 2009; Hassan et al., 2018). Sodium chloride (NaCl) accumulation induces metabolic disturbances that reduce germination rate and seedling emergence (Murillo-Amador et al., 2002; Janmohammadi et al., 2008). Excess salinity slows germination, while sub-lethal salinity levels may trigger dormancy (Khān & Weber, 2008), reflecting the plant's attempt to minimize damage through metabolic adjustments (Allen, 1994) or by adopting inherent tolerance mechanisms (Rumbaugh et al., 1993). Studies have shown that salinity negatively affects multiple germination parameters in maize, including germination rate, mean germination time (MGT), vigor index, and seedling growth (Kaya et al., 2008; Bybordi, 2010; Hassan et al., 2018). The detrimental effects extend to root and shoot elongation, with reductions in plumule length, shoot length, and seed vigor recorded at rates of up to 78–95% (Khodarahmpour et al., 2012). Additionally, germination index tends to be higher in small-sized seeds compared to larger ones under saline conditions, suggesting morphological influences on stress sensitivity. Despite these challenges, certain pre-sowing treatments like seed priming with NaCl have shown promise in enhancing germination and seedling establishment under saline environments (Farooq et al., 2006). Nevertheless, the severity of germination inhibition varies with salinity intensity and the specific genotype, underlining the importance of selecting salt-tolerant cultivars and optimizing management practices to mitigate early stage salinity stress (Djanaguiraman et al., 2003; Kamrul et al., 2018).

4.4.1.2 Vegetative Stage

Salinity stress poses a significant limitation to the vegetative development of maize, primarily by impairing critical physiological and morphological processes. Elevated soil salinity disrupts leaf initiation, expansion, and internode elongation, ultimately reducing overall plant growth and promoting premature leaf abscission (Akram et al., 2010; Qu et al., 2012). One of the key physiological consequences of salinity-induced growth reduction is apoplastic acidification, which alters the activity of cell wall-modifying enzymes essential for cell expansion. In maize, salt stress has been shown to inhibit cell division and reduce both the number and elongation rate of expanding cells, leading to a marked decline in leaf and shoot growth. Numerous studies have consistently reported that shoot growth is particularly sensitive to salinity stress in maize (El Sayed & El Sayed, 2011; Konuşkan et al., 2017; Ahmed et al., 2023). In addition to shoot inhibition, salinity significantly reduces leaf and stem biomass, leaf area, and both shoot and root dry weights (Evlagon et al., 1992; Ashrafuzzaman et al., 2002; Niu et al., 2012). These reductions are not only structural but are also linked to disruptions in key metabolic processes. For instance, salt stress adversely affects the activity of enzymes involved in nitrogen assimilation, thereby lowering nitrogen uptake and metabolism in maize and other cereal crops (Munns et al., 2006). In maize, shoot tissues exhibit greater sensitivity to salt stress compared to roots (Munns 2002). Salinity suppresses shoot development by inhibiting leaf initiation and expansion, and by retarding internode elongation, ultimately resulting in reduced overall shoot growth (Rios-Gonzalez et al., 2002). At the cellular level, one of the earliest responses to salt stress is membrane destabilization, primarily due to the displacement of calcium ions (Ca^{2+}) from plasma membrane binding sites (Cramer et al., 1998). Salinity impairs the proton pumping mechanisms that maintain cell wall acidification, a process essential for cell expansion and elongation, and dependent on ATP-driven H^+ efflux across the plasma membrane (Spanswick, 1981; Hager, 2003). Disruption of this acidification process leads to a significant inhibition of cell enlargement and growth.

Photosynthesis, a key determinant of biomass accumulation and yield, is also adversely affected by salinity. One of the major impacts is the significant reduction in chlorophyll content, a vital pigment for light absorption and energy conversion, along with decreased carotenoid levels (Tuna et al., 2013; Methela et al., 2017; Zahra et al., 2018). Salinity-induced stress leads to a decline in chlorophyll fluorescence efficiency (Fv/Fm), a reliable indicator of photosystem II functionality. A marked reduction in Fv/Fm has been reported under salt concentrations exceeding 10 $dS{\cdot}m^{-1}$, indicating substantial photoinhibition in maize seedlings (Deng et al., 2010; Stefanov et al 2021).

The cumulative effects of reduced leaf growth, premature leaf senescence, and impaired pigment synthesis culminate in diminished photosynthetic capacity. Moreover, prolonged salt stress exacerbates ion toxicity, disrupts membrane integrity, and induces stomatal closure—further restricting gas exchange and carbon assimilation (Omoto et al., 2012). Carbon fixation in maize is highly dependent on the proper functioning of photosynthetic enzymes, stomatal conductance, pigment stability, and structural integrity of the photosynthetic apparatus—all of which are negatively influenced by salinity stress (Gong et al., 2011; Qu et al., 2012). Ion homeostasis is another critical aspect affected by salinity. Under saline conditions, there is a pronounced accumulation of sodium (Na^+) and chloride (Cl^-) ions in the aerial parts of maize plants, with Cl^- often exceeding Na^+ levels (Turan et al., 2009; Carpici et al., 2010; Niu et al., 2012). Meanwhile, salinity hampers the translocation of potassium (K^+) from roots to shoots, leading to a reduced K^+ concentration in shoot tissues relative to roots (Kaya et al., 2010). These ionic imbalances—shaped by genotype, developmental stage, duration of stress, and soil fertility—further disrupt physiological processes, contributing to growth inhibition and reduced productivity. Salt stress profoundly alters plant physiological processes by disrupting water relations, ion homeostasis, nitrogen and carbon assimilation pathways, and protein biosynthesis (Khajeh-Hosseini et al., 2003; Hassan et al., 2018). The extent of growth inhibition under saline conditions is influenced by multiple factors, including the concentration and type of salts present in the soil, as well as the sensitivity of specific plant tissues to these salts (Munns & Tester, 2008). Among the primary effects of salinity stress is the accumulation of sodium (Na^+) and chloride (Cl^-) ions, which compromise the integrity of root cell membranes. This damage alters membrane selectivity, disrupts potassium (K^+) homeostasis, and interferes with nutrient uptake. The imbalance in ionic composition is further exacerbated by a reduction in the activity of key enzymes involved in nitrogen metabolism, such as nitrate reductase, thereby limiting nitrogen assimilation (Chattha et al., 2023). Additionally, salinity impairs the photosynthetic machinery, notably through inhibition of photosystem II and accelerated degradation of chlorophyll, resulting in diminished photosynthetic efficiency and overall plant vigor (Hasegawa et al., 2000; Zahra et al., 2018).

4.4.1.3 Reproductive Stage

Salinity is a major abiotic constraint that substantially hampers crop growth and productivity worldwide (Hasan et al., 2019). In maize, reproductive development is particularly vulnerable to salt stress, with GY being significantly diminished due to reductions in both grain number and grain weight (Schubert et al., 2009). These yield penalties are primarily attributed to impaired photosynthetic activity and reduced photoassimilate production under saline conditions (Hiyane et al., 2010; Hu et al., 2022). Salt stress limits biomass accumulation and disrupts the efficient translocation of assimilates from source tissues (leaves) to sink tissues (developing grains), ultimately leading to poor grain setting and incomplete grain filling (Lohaus et al., 2000).

Empirical studies have documented substantial yield reductions in maize under salinity stress. For instance, Sharif et al. (1999) reported a 22% decrease in GY at an electrical conductivity (EC) of 20 $dS{\cdot}m^{-1}$. Similarly, Cucci et al. (2019) observed a 34% decline in GY when soil salinity increased from 3.8 to 7.4 $dS{\cdot}m^{-1}$. These variations in yield loss are influenced by several factors, including maize genotype, developmental stage at the onset of stress, and environmental conditions during the growing season. Collectively, these findings underscore the critical need for salt-tolerant cultivars and agronomic strategies that mitigate the impact of salinity during the reproductive stage to secure maize productivity in salt-affected regions.

4.4.2 Mechanisms of Salt Stress

Maize is generally considered moderately sensitive to salinity stress (Zörb et al., 2004). According to Hasanuzzaman et al. (2013), a crop is classified as moderately tolerant if it maintains normal growth and development in soils with an electrical conductivity of the saturated extract (ECe) ranging from 3 to 6 $dS{\cdot}m^{-1}$. Salt stress in plants typically progresses through two distinct phases: an initial osmotic phase, followed by a phase characterized by specific ion toxicity (Munns,2008). Consequently, plant tolerance mechanisms are also dichotomized into strategies that address each of these phases.

During the osmotic phase, salinity reduces the soil's water potential, making water uptake more difficult. Plants respond by synthesizing and accumulating compatible solutes—such as proline, GB, and sugars—which help maintain cellular turgor by lowering the tissue water potential (Meloni et al., 2001). These osmoprotectants also play roles in stabilizing proteins and membranes under stress conditions. In the ionic phase, long-term exposure to high concentrations of Na^+ and Cl^- leads to toxicity and nutrient imbalances. Tolerance mechanisms at this stage include ion exclusion, where root tissues limit the uptake or translocation of toxic ions, and tissue tolerance, where cells compartmentalize ions into vacuoles to prevent cytosolic damage (Munns & Tester, 2008). Additionally, plants can redistribute Na^+ ions from shoots back to roots via the phloem as a protective strategy to minimize ion accumulation in photosynthetically active tissues (Tester & Davenport, 2003). The relative contribution of these mechanisms varies significantly between salt-sensitive and salt-tolerant species or genotypes, indicating that tolerance is a multifaceted trait shaped by complex physiological, biochemical, and genetic factors (Munns & Tester, 2008). Understanding and harnessing these mechanisms is key to improving salt stress resilience in maize and other cereal crops. Maize plants have evolved a suite of adaptive mechanisms to withstand the deleterious effects of saline environments. These strategies function in an integrated manner to mitigate osmotic and ionic stresses, restore cellular homeostasis, and sustain growth under saline conditions. The principal tolerance mechanisms include:

4.4.2.1 Ion Homeostasis and Compartmentalization

To mitigate ion toxicity, maize plants employ selective ion transport systems that exclude excess sodium (Na^+) and chloride (Cl^-) from entering the cytoplasm. Instead, these ions are compartmentalized into vacuoles to maintain cytosolic ion balance. Additionally, ion transporters such as SOS1 (Salt Overly Sensitive 1), HKT (High-Affinity Potassium Transporters), and NHX (Na^+/H^+ antiporters) play crucial roles in maintaining K^+/Na^+ homeostasis, thereby preserving membrane integrity and enzyme function.

4.4.2.2 Biosynthesis of Osmoprotectants and Compatible Solutes

Under salt-induced osmotic stress, maize synthesizes various osmoprotectants—including proline, GB, sugars, and polyols—that lower cellular osmotic potential and help maintain turgor pressure. These molecules also stabilize proteins, membranes, and enzymes, contributing to cellular resilience under stress conditions.

4.4.2.3 Biosynthesis of Antioxidants and Enzyme Activation

Salinity stress induces the overproduction of ROS, which can damage cellular components. Maize counteracts this oxidative damage through enhanced activity of antioxidant enzymes such as SOD, CAT, and POD, along with the synthesis of non-enzymatic antioxidants like ascorbate and glutathione. These defense systems collectively scavenge ROS and protect cellular structures.

4.4.2.4 Hormonal Regulation

Phytohormones play pivotal roles in modulating stress responses. Salinity alters the endogenous levels of hormones such as ABA, SA, ET, and cytokinins (CKs). ABA, in particular, accumulates rapidly under salt stress, promoting stomatal closure to reduce transpiration and activating stress-responsive genes. A fine-tuned hormonal balance is essential for coordinating growth and stress adaptation.

4.4.2.5 Apoplastic Acidification

Salt stress can modulate the apoplastic pH through H^+-ATPase-driven proton extrusion across the plasma membrane. Acidification of the apoplast loosens the cell wall and facilitates cell expansion, which can be a compensatory growth response under moderate salinity. This process is also associated with the activation of expansions and other cell wall-modifying enzymes.

4.4.2.6 Synthesis of Polyamines

Polyamines (e.g., putrescine, spermidine, and spermine) are low-molecular-weight polycations that accumulate in maize under salt stress. These compounds stabilize membranes, scavenge ROS, and interact with signaling molecules to regulate stress responses, including gene expression and ion channel activity.

4.4.2.7 Generation of Nitric Oxide (NO)

NO functions as a signaling molecule that modulates salt stress responses by influencing gene expression, ion transport, and antioxidant activity. It interacts with ROS to form peroxynitrite, triggering defense pathways and enhancing salt tolerance through post-translational modifications of proteins.

4.4.3 Strategies to Mitigate the Impacts of Salinity Stress

There is no doubt that salinity is a major constraint to global crop productivity, significantly limiting plant growth and yield. In response, several strategies have been developed and practiced to mitigate the adverse effects of salt stress and enhance the salt tolerance of crops. Among these, the exogenous application of various plant growth regulators and signaling molecules has emerged as a promising approach for improving plant performance under saline conditions. These strategies include the external application of phytohormones (e.g., ABA, SA, jasmonic acid, and brassinosteroids (BRs)), amino acids (such as proline and glutamate), sugars (e.g., trehalose, sucrose), osmolytes (e.g., GB), and other protective compounds that function as signaling agents or metabolic stabilizers. These exogenous treatments help enhance stress perception, modulate gene expression, improve osmotic adjustment, and boost antioxidant defense systems—ultimately alleviating the physiological and biochemical damage caused by salinity. Collectively, these interventions not only improve germination, root and shoot development, and photosynthetic efficiency but also contribute to greater ion homeostasis and membrane stability. As shown in Figure 4.2, these compounds act through various interconnected pathways to reinforce crop resilience under salt stress conditions.

4.4.3.1 Selection of Salt-Tolerant Genotypes

To enhance maize productivity in salt-affected regions, the development and deployment of salt-tolerant genotypes, alongside optimized agronomic practices, are of paramount importance. Genetic variability among maize genotypes offers a valuable opportunity for improving resilience to salinity stress, as different cultivars exhibit varying degrees of tolerance (Akram et al., 2010). Identifying and promoting relatively salt-tolerant genotypes that can maintain growth and yield under saline conditions is a vital strategy for sustainable production. The selection of salt-tolerant maize varieties not only ensures better germination and early seedling establishment but also enhances physiological and biochemical traits essential for stress mitigation. Hence, genotype screening and breeding for salt tolerance should be prioritized as a fundamental component of crop improvement programs targeted at saline-prone environments (Gobade et al., 2025). When combined with location-specific production technologies—such as soil amendments, irrigation management, and nutrient optimization—tolerant genotypes can significantly contribute to improving maize yields in marginal lands impacted by salinity.

4.4.3.2 Seed Priming

Achieving optimal crop stand establishment is critical for successful maize production, and this is largely dependent on the seed's ability to germinate effectively under field conditions. However, salinity poses a major challenge by significantly delaying and reducing seed germination due to osmotic stress, which limits water uptake and disrupts early metabolic processes in germinating seeds (Hasegawa et al., 2000; Farsiani & Ghobadi, 2009). Among the various strategies developed to counteract this issue, seed priming has emerged as a practical and efficient method for improving germination performance and early seedling vigor in saline environments (Table 4.1).

Several studies have demonstrated that hydropriming maize seeds under saline conditions enhances germination rate, percentage, germination index, seedling vigor index, and seedling length (Ashraf & Rauf, 2001; Janmohammadi et al., 2008). In addition to hydropriming, osmopriming with salts such as sodium chloride (NaCl), potassium chloride (KCl), and calcium chloride ($CaCl_2$) has been shown to mitigate the negative effects of salinity, promoting stronger seedling establishment and growth traits (Ashraf & Rauf, 2001; Bakht et al., 2011; Choudhary et al., 2021). Hormonal priming has also proven effective in improving maize germination and seedling development under salt stress. For example, priming with chloroethyl phosphonic acid was found to increase seedling biomass in saline conditions (Carvalho et al., 2011), while 28-homobrassinolide enhanced seedling growth by stimulating antioxidant enzyme activity, reducing lipid peroxidation, and increasing protein synthesis (Arora et al., 2008). SA priming has also been reported to support seedling development under salt stress by modulating stress-responsive physiological pathways (Khodary, 2004). Other effective priming agents include hydrogen peroxide, which boosts germination through increased activity of antioxidative enzymes such as APX, CAT, and guaiacol peroxidase (GPX) (Gondim et al., 2010), and vitamin-based priming agents such as ascorbic acid, thiamin, and pyridoxine, which improve seedling growth and yield under saline stress (Ahmed-Hamad & Monsaly, 1998). Furthermore, seed priming with 24-epibrassinolide (EBL) under varying salinity levels significantly enhanced multiple physiological and biochemical parameters, including root length, root surface area, stem diameter, relative leaf water content, chlorophyll content, PSII photochemical efficiency (Fv/Fm), net photosynthetic rate, transpiration rate, and nutrient uptake (Zn and K), while reducing Na^+ accumulation and Na^+/K^+ ratio in maize seedlings (Ahmad et al., 2023).

4.4.3.3 Organic Matter Management

The application of organic amendments has emerged as a practical and sustainable strategy to mitigate salinity stress in agricultural soils. Incorporating organic matter into saline soils enhances both their physical and chemical properties, thereby improving soil health and supporting crop productivity. Organic amendments—such as compost, vermicompost, farmyard manure, cow dung, and other organic wastes—not only reduce soil salinity but also enhance the nutrient availability and ionic balance necessary for optimal plant growth. These amendments contribute to improved maize performance under saline conditions by increasing soil organic carbon content and enhancing cation exchange capacity (CEC), which, in turn, helps reduce the accumulation of toxic sodium (Na^+) ions in the root zone. Additionally, organic amendments enrich the soil with essential nutrients, such as total nitrogen (N), available phosphorus (P), exchangeable potassium (K^+), calcium (Ca^{2+}), and magnesium (Mg^{2+}). This nutrient enrichment and ion balancing effect collectively promote better growth, physiological function, and yield of maize under salt-stressed environments.

4.4.3.4 Management of Nutrients

Salinity stress adversely affects maize growth and development by limiting the uptake of essential macro- and micronutrients such as nitrogen (N), potassium (K^+), calcium (Ca^{2+}), magnesium (Mg^{2+}), iron (Fe), and zinc (Zn) (Karimi et al., 2005; Gunes et al., 2007; Turan et al., 2009; Kaya et al., 2010; Qu et al., 2012). Nutrient deficiencies in salt-stressed plants primarily arise due to ionic imbalances and competitive inhibition, where the excessive accumulation of sodium (Na^+) and chloride (Cl^-) interferes with the uptake of beneficial cationic (NH_4^+, K^+, Ca^{2+}, Mg^{2+}, Zn^{2+}) and anionic (NO_3^-, PO_4^{3-}, SO_4^{2-}, Fe^{2+}) ions (Islam et al., 2011). Significant reductions in the concentrations of K^+, Ca^{2+}, and Mg^{2+} in both leaves and roots have been reported in maize under combined salt and nutrient deficiency stress compared to salt stress alone (Qu et al., 2012; Shahzad et al., 2012). The accumulation of Na^+ and Cl^- is notably higher under salinity and is further intensified in conditions of potassium deficiency. Consequently, critical nutrient ratios such as K^+/Na^+, Ca^{2+}/Na^+, and Mg^{2+}/Na^+ decline drastically under dual stress conditions, impairing physiological processes and metabolic stability.

To alleviate the negative effects of salinity, the application of essential nutrient elements—either as elemental forms or fertilizer formulations—has proven effective. For instance, nitrogen supplementation mitigates salinity-induced growth inhibition and enhances nutrient uptake, photosynthetic efficiency, and GY in maize (Gadalla et al., 2007). N application also improves physiological attributes such as leaf area index, crop growth rate, and net assimilation rate under salt-stressed conditions (Akram et al., 2010). Additionally, it modulates antioxidant defense by upregulating the activities of key antioxidant enzymes, including CAT, glutathione reductase (GR), and GST, while decreasing the activity of POD and SOD (Rios-Gonzalez et al., 2002). Thus, not only the quantity but also the source and timing of nitrogen application play a critical role in mitigating salt stress in maize. The application of macro- and micronutrients plays a vital role in mitigating salinity stress and improving maize performance in salt-affected soils. Potassium (K) application significantly enhances yield-related traits and overall GY in maize under saline conditions (Maqsood et al., 2008). This improvement is attributed to enhanced water uptake, maintenance of cell turgor, improved mineral nutrition, higher leaf area index, increased crop growth rate, and the regulation of Na^+ uptake through osmoregulation (Akram et al., 2010). Elevated Na^+ levels in saline soils often result in a high Na^+/K^+ ratio, impairing plant metabolism. However, exogenous K application effectively reduces this ratio and mitigates the toxic effects of Na^+ (Akram et al., 2010; Hussain et al., 2015). Furthermore, combined soil application of potassium sulfate and diammonium phosphate (DAP) significantly improved maize growth and yield under salinity stress (Hussain et al., 2015). In addition to soil amendments, foliar applications also show promising effects. Foliar application of potassium chloride (KCl) and thidiazuron notably improved maize growth under salinity. Thidiazuron, in particular, proved more effective than K fertilization alone, as it significantly increased K^+ uptake, chlorophyll content, total carbohydrates, and soluble sugars, thereby enhancing salt tolerance (El-Sharkawy et al., 2017). Among the micronutrients, boron (B) plays a crucial role in stabilizing the cell wall and membrane structure (Sezer, 2014; El-Sharkawy et al., 2017). B application in salt-stressed maize improved membrane integrity by increasing K^+ concentrations and alleviating the detrimental effects of Na^+ and Cl^-, leading to enhanced growth and productivity (Bastías et al., 2004). Similarly, silicon (Si) has been demonstrated to improve photosynthetic machinery in maize under saline stress by enhancing electron transport chain efficiency (Khan et al., 2016). Si application contributes to enhanced salt tolerance by reducing oxidative and osmotic stress, while promoting physiological attributes such as water use efficiency and pigment synthesis (Khan et al., 2019). Moreover, combined application of B and Si showed synergistic effects, improving maize growth and physiological performance under salinity by increasing the accumulation of proline and total soluble sugars (Salim, 2014). Other foliar treatments have also shown efficacy. For example, foliar-applied iron (Fe) improved the nutritional status of salt-stressed maize seedlings (Salama et al., 1996). Additionally, hydrogen peroxide (H_2O_2) foliar application has been reported to enhance salt tolerance in maize by stimulating antioxidant defense enzymes such as APX, GPX, SOD, and CAT (Gondim et al., 2012).

4.4.3.5 Fresh Water Irrigation

Effective irrigation strategies are essential for mitigating the adverse impacts of salinity stress on maize growth and yield. Alternating saline and fresh water during different growth stages has been shown to improve crop performance under saline conditions. Irrigation with fresh water at the jointing, heading, and grain-filling stages of maize increased GY by 81.16% compared to continuous saline water irrigation at those stages. Even a two-time saline and one-time fresh water schedule enhanced yield by 35.39% (Sang et al., 2020). Alternating irrigation of saline and fresh water not only reduced soil salinity but also improved GY in maize (Wang et al., 2016). Specifically, fresh water irrigation during the seedling stage effectively promoted salt leaching and enhanced yield (Zhang et al., 2018), while application at the heading stage helped recover earlier stress, improving growth and water use efficiency (Sun et al., 2014). The concept of alternating fresh and saline water applications has also proven effective in other crops. For example, applying fresh water during salt-sensitive stages and saline water during more tolerant stages increased the utilization efficiency of saline water (Gawad et al., 2005). A fresh-to-saline water ratio of 80:20 yielded tomato crops comparable to those grown under 100% fresh water irrigation (Malash et al., 2005). Similarly, alternate irrigation resulted in higher wheat and mustard yields than mixed irrigation with the same total salt content (Minhas, 1996). In wheat and barley, saline water could replace fresh water with minimal yield loss under deficit irrigation strategies, even at EC levels as high as 9 $dS{\cdot}m^{-1}$ (Hamdy 2005).

Alternate irrigation strategies effectively regulate soil salinity and reduce salt toxicity. Specifically, alternating fresh and saline water during the seedling, jointing, tasseling, and post-tasseling stages significantly enhanced maize yield and alleviated salinity stress (Zhu et al., 2007; Huang et al., 2019). Brackish water irrigation, when applied judiciously, also helped alleviate salinity stress by improving K^+ uptake and limiting Na^+ accumulation in certain maize genotypes (Tedeschi & Menenti, 2002). However, the irrigation frequency influenced salt accumulation in soil under brackish water use. Saline water irrigation modifies the soil's physicochemical properties, including permeability, capillary rise, and water retention, which may lead to salt accumulation and reduced crop productivity (Huang et al., 2011; Ahmed et al., 2012; Chen et al., 2017; He et al., 2017). Combined irrigation treatments—such as applying 80 mm each of fresh, brackish, and saline water—have been reported to reduce soil salinity and increase winter wheat yield (Wang et al., 2019). Moreover, frequent saline water irrigation (e.g., daily intervals) under controlled conditions has also been shown to enhance tomato growth and yield (Rahil et al., 2013). These findings underscore the importance of tailoring irrigation strategies to both crop sensitivity and water quality to sustain productivity in saline-prone areas.

4.4.3.6 Exogenous Application of Phytohormones

The exogenous application of phytohormones has been shown to mitigate the negative impacts of salt stress on maize plants, enhancing growth and productivity. Specifically, the application of IAA has been reported to improve the growth and yield of salt-stressed maize by promoting the uptake of essential nutrients and reducing the uptake of sodium (Na^+) (Darra & Saxena, 1973). In a study on maize grown under saline conditions, the combined application of inorganic nutrients and IAA resulted in increased concentrations of phosphorus (P), calcium (Ca), and magnesium (Mg), while simultaneously lowering Na^+ levels. This improvement in plant growth was associated with higher concentrations of photosynthetic pigments, a more favorable K^+/Na^+ ratio, enhanced activities of antioxidant enzymes such as CAT and SOD, and reduced membrane permeability under saline conditions. Furthermore, IAA was found to mitigate Na^+ and chloride (Cl^-) toxicity in maize, thereby promoting higher K^+ and Ca^{2+} concentrations and maintaining optimal membrane permeability (Darra & Saxena, 1973). These findings align with other studies, such as those on wheat, where exogenous IAA alleviated salt stress and improved seed quality, growth, and yield (Akbari et al., 2007).

Gibberellic acid (GA_3), another important phytohormone, is known to improve the growth, yield, and quality of crops, including maize (Islam et al., 2023). Salinity stress has been shown to reduce the endogenous levels of GA_3 in plants (Elhakem, 2020), contributing to the oxidative stress and ionic toxicity that inhibit plant growth. However, the exogenous application of GA_3 has proven to counteract these negative effects by enhancing physiological and biochemical traits, including macro- and micronutrient concentrations. In maize, GA_3 application under NaCl-induced saline stress alleviated the adverse effects of Na^+ and Cl^-, improving growth and nutritional status. It also increased the concentration of proline, which helped maintain membrane integrity and enhanced salt tolerance (Tuna et al., 2008). Foliar application of GA_3 in salt-stressed maize has been shown to increase nutrient uptake and reduce the concentrations of Na^+ and Cl^-, thereby improving plant growth (Rashad & Hussien, 2014). Notably, GA_3 was found to be more effective in enhancing salinity tolerance in maize than SA.

CKs play a central role in regulating plant metabolism and development, and any disruption in their synthesis or homeostasis can lead to tissue senescence or plant death (Rijavec et al., 2011). CKs are known to enhance plant adaptation to environmental stresses, including salinity, by improving physiological tolerance mechanisms (Iqbal et al., 2006; Ha et al., 2012; Taiz et al., 2015). Exogenous application of CKs has been reported to support various critical processes involved in plant growth and development. In maize, foliar application of CKs significantly alleviated the detrimental effects of salt stress, leading to improved vegetative growth, yield-related traits, and GY (Davani et al., 2017). Specifically, kinetin has been shown to improve salt tolerance in maize by regulating shoot chloride exclusion (Chakrabarti & Mukherji, 2003). Foliar spraying of kinetin in combination with IAA further ameliorated the effects of NaCl-induced stress by increasing membrane stability, enhancing K^+ and Ca^{2+} uptake, and reducing Na^+ accumulation (Kaya et al., 2010). ET, another stress-responsive phytohormone, is synthesized in greater quantities under salinity stress. ET has been shown to aid in maintaining ionic balance, particularly the Na^+/K^+ ratio, thereby contributing to enhanced salt tolerance (Jiang et al., 2013). Recent studies suggest that ET improves salinity tolerance in maize by modulating the activity of enzymes involved in polyamine catabolism, which are critical for cellular homeostasis during stress (Freitas et al., 2018). ABA, a well-established stress-signaling phytohormone, plays a vital role in mitigating salt-induced damage.

ABA signaling helps plants survive under saline and drought conditions by regulating physiological and biochemical responses (Keskin et al., 2010). Salt stress triggers an increase in endogenous ABA levels, particularly in salt-tolerant maize genotypes (Zhang et al., 2006; Zörb et al., 2013). Elevated ABA concentrations under saline conditions h

ave been associated with reduced leaf expansion and restricted accumulation of toxic ions such as Na^+ and Cl^- in maize leaves (Cabot et al., 2009), thereby contributing to stress adaptation. BRs play a pivotal role in mitigating the adverse effects of environmental stresses, including salinity-induced oxidative damage (Cao et al., 2005). Under salt stress conditions, BRs have been widely studied for their influence on plant growth and antioxidant defense mechanisms (Arora et al., 2008; Ali et al., 2008, 2008; Rady, 2011). BRs are involved in regulating several key physiological processes such as cell division, cell elongation, xylem differentiation, stem elongation, and root development (Cao et al., 2005; Kartal et al., 2009). Their exogenous application has been shown to restore seedling growth in maize affected by salinity stress (He et al., 1991). Moreover, BRs enhance seed germination and seedling vigor under salt stress by modulating oxidative responses, thus improving the plant's stress tolerance (Wang et al., 2011). SA, a phenolic compound also known as 2-hydroxybenzoic acid, functions as a key regulatory phytohormone in plant defense against both biotic and abiotic stresses. SA mitigates salinity-induced damage by modulating physiological and biochemical processes. Under NaCl-induced salt stress, maize plants exhibit reduced growth, impaired photosynthetic pigment synthesis, disturbed water relations, increased electrolyte leakage, and decreased antioxidant enzyme activity. However, exogenous application of SA significantly reverses these negative effects, enhancing salt tolerance by restoring photosynthetic efficiency, water status, and antioxidant defense (Agami, 2013). SA has also been shown to improve nutrient uptake by reducing Na^+ and Cl^- accumulation in plant tissues (Gunes et al., 2007). Several studies corroborate the role of SA in enhancing maize tolerance to both salinity and osmotic stress (Tuna et al., 2007; Ismail, 2013). Therefore, the exogenous application of phytohormones such as BRs and SA serves as an effective strategy to enhance maize performance under saline conditions. These phytohormones contribute to improved nutrient uptake, osmotic adjustment, and activation of antioxidant defense systems, thereby promoting stress resilience and sustainable growth.

4.4.3.7 Exogenous Application of Osmoprotectants

Compatible osmolytes play a significant role in mitigating the adverse effects of salinity stress and enhancing plant growth and development. Among them, proline is a principal osmolyte synthesized by the hydrolysis of proteins under abiotic stress conditions, contributing to osmotic adjustment and cellular homeostasis. Numerous studies have demonstrated that salt stress reduces maize growth and yield by decreasing total nutrient uptake and promoting excessive accumulation of toxic ions such as Na^+ and Cl^- (Azevedo Neto et al., 2004; de Freitas et al., 2018). However, the exogenous application of proline has been shown to alleviate these negative effects by enhancing nutrient acquisition and reducing Na^+ accumulation in maize plants (Hasan et al., 2015; Alam et al., 2017). Hamd and Wakeel (1994) further confirmed that proline supplementation significantly increased salt tolerance by improving nutrient uptake under saline conditions. GB, another crucial osmoprotectant, is effective in counteracting the impacts of abiotic stresses, particularly salinity. Exogenous GB application has been reported to enhance growth parameters, water status, net photosynthesis, and yield in salt-stressed maize by improving stomatal conductance and photosystem II efficiency (Yang & Lu 2005). Moreover, GB supports salt tolerance by decreasing Na^+ and increasing K^+ concentrations in shoot tissues. GB also plays a pivotal role in maintaining enzymatic activity, regulating ROS homeostasis, and bolstering the antioxidant defense system under stress conditions (Sabagh et al., 2021).

In addition, hydrogen peroxide (H_2O_2), when applied exogenously in low concentrations, functions as a signaling molecule that initiates stress acclimation mechanisms. H_2O_2 has been found to enhance salt tolerance in maize by increasing the activity of key antioxidant enzymes such as CAT, GPX, APX, and SOD (Gondim et al., 2012). Pretreatment with H_2O_2 was shown to improve resistance in maize by facilitating salt acclimation responses (de Azevedo Neto et al., 2005).

Furthermore, the use of hydrogel polymers under saline irrigation conditions has shown promise in improving maize growth and morpho-physiological traits. This is primarily achieved through enhanced soil properties, which facilitate better water retention and nutrient availability under salt stress (El Sayed & El Sayed, 2011).

4.4.3.8 Applications of Arbuscular Mycorrhizal Fungi

Arbuscular mycorrhizal fungi (AMF) play a pivotal role in enhancing plant stress tolerance and promoting growth under abiotic stress conditions, including salinity. In maize, AMF have been shown to significantly improve photosynthetic efficiency and water use efficiency under saline environments (Hashem et al., 2019). AMF colonize plant roots by penetrating cortical cells and forming specialized structures such as arbuscules and vesicles. These structures facilitate the uptake of both macro- and micronutrients from the soil, thereby enhancing plant resilience to salinity. For instance, the AM fungus *Funneliformis mosseae* increased salt tolerance in maize by enhancing tissue density, facilitating Na^+ exclusion from leaves, modulating Na^+ distribution within plant organs, maintaining a favorable K^+:Na^+ ratio, and preserving organelle structure and function (Wang et al., 2021).

Additionally, AMF colonization has been associated with increased accumulation of soluble sugars in salt-stressed maize roots, which contributes to improved osmotic adjustment and overall salt tolerance (Kumar et al., 2016). This improvement is strongly linked to enhanced nutrient uptake, particularly the maintenance of higher K^+/Na^+ ratios. Moreover, AM symbiosis influences the expression of various stress-related plant genes. It modulates genes involved in proline biosynthesis, aquaporin production, and the synthesis of LEA proteins, which act as molecular chaperones (Porcel et al., 2012). Notably, maize encodes over 30 different aquaporin genes (Chaumont et al., 2001), and mycorrhizal colonization helps maintain water balance in plant tissues by regulating their expression. Although salt stress may reduce the rate of AMF colonization in maize roots (Hashem et al., 2019), mycorrhizal plants consistently exhibit greater tolerance to salinity compared to non-mycorrhizal counterparts (Porcel et al., 2012; Kumar et al., 2015; Dubey et al., 2020; Thind et al., 2022). This highlights the potential of AMF as a sustainable strategy to mitigate the impacts of salinity on maize cultivation.

4.4.3.9 Molecular Strategies

Molecular studies have significantly advanced our understanding of salt tolerance mechanisms in maize. Key research efforts have identified numerous genes and regulatory pathways involved in conferring salt stress resilience. For instance, Luo et al. (2019) reported the discovery of crucial genes related to salt tolerance, while Hussain et al. (2023) investigated the physiological, biochemical, and molecular responses of contrasting maize genotypes under saline conditions. GWAS have further elucidated the genetic basis of salt tolerance traits in maize (Zhao et al., 2023), and morphological classification has been used to differentiate maize genotypes based on their tolerance levels (Li et al., 2023).

Several genes have emerged as central players in maize salt tolerance, including *ZmPIF3*, *ZmHAK1*, *ZmNHX1*, *ZmSTG1*, and *Zm00001d053925*, each contributing to stress adaptation through diverse physiological and biochemical mechanisms (Zhao et al., 2023; Li et al., 2023; Han et al., 2023; Mei et al., 2023). In particular, Luo et al. (2021) utilized GWAS to identify salt tolerance markers and employed CRISPR/Cas9 to validate candidate gene functions, paving the way for targeted genetic improvement strategies. These genes regulate various processes such as ion transport, osmotic adjustment, stress signaling, and photosystem protection under saline conditions.

Maize counters salt stress by activating defense mechanisms, including osmolyte accumulation, antioxidant enzyme activity, and controlled ion transport (Li et al., 2023). Moreover, studies have revealed the critical roles of phytohormone signaling, calcium (Ca^{2+}) signaling, and TFs in orchestrating salt stress responses. The identification of salt tolerance QTLs and the development of salt-tolerant maize cultivars (Cao et al., 2023) underscore the potential of molecular breeding and biotechnological approaches to enhance crop resilience in saline environments. Overall, these findings highlight the importance of molecular insights in designing effective strategies for improving maize productivity under salt stress.

4.5 HEAT STRESS: PHYSIOLOGICAL, GENETIC, AND MOLECULAR INSIGHTS

Climate change is increasingly undermining global crop productivity, posing a significant threat to food security (Cairns et al., 2013; Farooq et al., 2022). Among the most affected are major staple crops such as maize (*Zea mays* L.), rice (*Oryza sativa* L.), and wheat (*Triticum aestivum* L.), which collectively support a large proportion of the global population's caloric intake. Multiple abiotic stressors driven by climate change—including rising average temperatures, intensified heatwaves, prolonged droughts, expanding soil salinization, nutrient imbalances, and the accumulation of toxic heavy metals—are contributing to this decline in crop performance (Dhankher & Foyer, 2018; Rivero et al., 2022).

Of these factors, increasing mean temperature has emerged as one of the most critical constraints on agricultural productivity, with numerous studies reporting its substantial negative effects on plant growth, development, and yield (Raza et al., 2021, 2023; Haider et al., 2022; Sharma et al., 2022). According to NASA's Goddard

Institute for Space Studies (GISTEMP, 2023), the global mean surface temperature has risen by approximately 1.1°C since the pre-industrial era (1880), with a marked increase in the frequency and intensity of heatwaves observed in recent decades. This trend is particularly detrimental to crop production in rainfed agroecosystems, where water scarcity and temperature stress act synergistically to reduce productivity (Chakraborty et al., 2019).

Maize (*Zea mays* L.) is the most extensively cultivated C_4 crop, covering approximately 197 million hectares globally, and ranks as the third most important cereal in terms of contribution to global food security (FAOStat, 2021). According to the U.S. Department of Agriculture (USDA), global maize production reached over 1,150 million metric tons during the 2022/2023 marketing year. Despite its agronomic significance and wide adaptability, maize is increasingly exposed to frequent and intense episodes of heat stress, which severely affect its physiological functioning across morphological, physiological, biochemical, and molecular levels—ultimately compromising growth and yield (Wang et al., 2021a).

A growing body of evidence underscores the detrimental impacts of heatwaves on maize production in various regions, including Europe (Hawkins et al., 2013), the United States (Djalovic et al., 2024), sub-Saharan Africa (Lobell et al., 2011), and China (Tao et al., 2016; Gao et al., 2021). In response, enhancing heat tolerance in maize has become a priority research area for plant biologists, breeders, and crop scientists worldwide (Rahman et al., 2015; Iqbal et al., 2020).

In recent years, crop growth simulation models have been widely employed to predict the effects of climate-induced heat stress on maize productivity (Rezaei et al., 2015; Gabaldón-Leal et al., 2016). However, many of these models lack specific calibration for extreme heat events or do not incorporate mechanistic frameworks to accurately simulate stress at critical developmental stages. As a result, they may under- or overestimate the impacts of heat stress on maize yield. For instance, while elevated temperatures during the flowering stage are well documented to cause substantial yield reductions, similar heat conditions during earlier vegetative stages may have limited or negligible effects (Lizaso et al., 2018;). Specifically, Lizaso et al. (2018) reported that maize growth and GY were unaffected by heat stress occurring before the ninth leaf (V9) stage. Recent studies have further clarified the sensitivity of different developmental phases to thermal stress, with particular emphasis on reproductive stages, which are especially vulnerable. Both short- and long-duration heat stress have been shown to impair reproductive processes and significantly reduce the harvest index in cereals (Prasad et al., 2017). In maize, tassel and ear initiation occurs around the sixth leaf collar stage (V6), a critical window during which heat stress can lead to malformed or underdeveloped tassels. Severe heat exposure at this stage can result in "tassel blast" (TSBL)—a condition characterized by desiccation and premature death of tassels—ultimately diminishing pollination success and yield potential.

Reproductive tissue damage due to heat stress, particularly in the form of TSBL, has been shown to significantly reduce viable pollen production (McNellie et al., 2018). Beyond pollen formation, the entire process of pollen development is highly sensitive to elevated temperatures. Heat stress can impair pollen tube growth and diminish stigma receptivity due to silk desiccation, ultimately disrupting successful fertilization and reducing seed set (Zinn et al., 2010; Lohani et al., 2020). These disruptions collectively contribute to decreased GY under high-temperature conditions. Furthermore, the synchronized development of ear shoots is essential for silking and successful grain filling. Heat stress occurring during the pre-anthesis phase can impede this developmental coordination, adversely affecting yield potential (Suwa et al., 2010). During the grain-filling stage, elevated temperatures further exacerbate yield loss by increasing respiratory carbon losses and disrupting the translocation of assimilates from source tissues to developing kernels (Yang et al., 2018). Additionally, heat stress shortens the duration of the active grain-filling period (Yang et al., 2017), leading to reduced kernel weight and compromised grain quality (Yang et al., 2018; Chukwudi et al., 2021). Given the multifaceted impacts of heat stress across key developmental stages—particularly reproductive and grain-filling phases—it is imperative to deepen our understanding of the underlying molecular, biochemical, and physiological mechanisms. Such knowledge is essential to elucidate the complex plant responses that impair maize growth, development, and productivity under heat stress (Table 4.4).

TABLE 4.4
Physiological and Genetic Responses of Maize to Heat Stress

Response Type	Trait/Parameter Affected	Observed Change under Heat Stress	Physiological/Genetic Implication	References
Physiological	Seed germination and seedling vigor	Decreased	Poor crop establishment and reduced early vigor	Tas (2022)
	Root elongation and leaf expansion	Impaired	Reduced water and nutrient uptake; limited photosynthetic capacity	Tas (2022)
	Stomatal conductance	Altered	Disrupted gas exchange and evaporative cooling; lower tissue water potential	Urban et al. (2017)
	Leaf temperature and chlorophyll content	Increased leaf temperature; decreased chlorophyll a/b ratio	Decline in photosynthetic efficiency; early senescence	Coşkun et al. (2011)
	ROS accumulation and membrane integrity	Elevated ROS levels; membrane damage	Oxidative stress leading to cellular damage and protein degradation	Tas (2022)
	Total phenol content (TPC)	Reduced	Weakened antioxidant defense system	Tas (2022)
	Anthesis-silking interval (ASI)	Prolonged	Impaired synchronization of male and female flowering; low kernel set	Karim et al. (1999); Thayil et al. (2020)
	Reproductive traits: pollen viability, stigma receptivity	Declined	Compromised fertilization and grain formation	Alam et al. (2017); Noor et al. (2019a)

(Continued)

TABLE 4.4 *(Continued)*
Physiological and Genetic Responses of Maize to Heat Stress

Response Type	Trait/Parameter Affected	Observed Change under Heat Stress	Physiological/Genetic Implication	References
	Delayed senescence ("stay-green")	Enhanced in tolerant genotypes	Marker of thermotolerance; associated with sustained photosynthesis	Noor et al. (2019a)
	Grain filling and assimilate accumulation	Shortened duration; reduced accumulation	Lower grain yield and compromised grain quality	Coşkun et al. (2011)
	Yield-associated morphological traits	Altered ear aspect, husk cover, ears per plant	Strong correlation with yield stability under heat stress	Nelimor et al. (2019)
Genetic	Gene expression profile	Over 5,000 genes differentially expressed	Indicates polygenic and multilayered heat stress response	Joshi et al. (2021)
	QTLs for ASI, pollen viability, tassel fertility	Identified	Useful for breeding heat-tolerant varieties	Parrado et al. (2021)
	Heat shock transcription factors (HSFs)	Upregulated (e.g., HSFA1, HSF3, HSF4, and HSF6)	Induce expression of heat shock proteins (HSPs); enhance thermotolerance	Lin et al. (2011)
	Transgenic expression (e.g., OsMYB55 overexpression)	Improved biomass and stress tolerance	Enhanced performance under heat and drought stress	Casaretto et al. (2016)
	PGD3 enzyme and starch synthesis	Heat-sensitive; engineered substitutes (WPGD1, WPGD2) function under heat	Restores kernel starch accumulation and improves yield under high night temperatures	Ribeiro et al. (2020); Spielbauer et al. (2013)
	SNP diversity and GWAS	170,098 SNPs identified; trait-associated alleles discovered	Enables precision breeding using marker-assisted selection	Ahmed et al. (2022)
	Genome editing (e.g., ARGOS8 and PSY1 via CRISPR-Cas9)	Edited genes enhanced drought/heat tolerance	Demonstrates potential for engineering complex traits like thermotolerance	Shi et al. (2017); Zhu et al. (2016)
	Linkage disequilibrium in landraces versus elite lines	Faster LD decay in landraces	Facilitates high-resolution mapping; identifies novel alleles for thermotolerance	Würschum et al. (2022)

This section of the chapter provides a comprehensive synthesis of current knowledge on the impacts of heat stress on maize. It further emphasizes the significance of stress-related, trait-based phenotyping and explores key molecular mechanisms that facilitate the identification of heat-tolerant genotypes. Collectively, these insights offer new opportunities for enhancing heat tolerance in maize through targeted breeding and biotechnological interventions.

4.5.1 Heat Stress in Nature

Advancements in agronomic practices and a significant expansion in cultivated area—from 106 million hectares in 1963 to approximately 197 million hectares at present—have contributed to a nearly fivefold increase in global maize production (Erenstein et al., 2022). Despite these gains, climate projections indicate that maize yields could decline by approximately 7.4% for every 1°C rise in average temperature (Zhao et al., 2017). Maize is a thermophilic crop with an optimum growing temperature range of 28–32°C, which is comparatively higher than that required for many other cereal crops (Arnold, 1974). However, heat stress begins to adversely affect physiological processes such as photosynthesis and vegetative growth when temperatures exceed 38°C (Crafts-Brandner & Salvucci, 2002). The reproductive stage is especially vulnerable to elevated temperatures, with prolonged exposure to temperatures above 35°C shown to severely impair reproductive development, reduce seed set, and lead to substantial losses in reproductive biomass and yield (Carter et al., 2016; Shao et al., 2021). Among reproductive structures, male floral organs (tassels) are more heat-sensitive than female counterparts (Dong et al., 2021). Heat stress at this stage interferes with key reproductive processes, including inhibition of anther dehiscence, reduced pollen viability, impaired pollen germination, and defective pollen tube growth (Figure 4.4).

A notable consequence of heat stress is the elongation of the ASI, often caused by delayed silk emergence and silk desiccation. This temporal mismatch between pollen shed and silk receptivity suppresses fertilization efficiency, resulting in increased kernel abortion and reduced kernel number per cob (Carberry et al., 1989; Bakhtavar et al., 2015; Hatfield & Prueger, 2015; Li & Howell, 2021). Additionally, heat stress occurring after pollination continues to pose a threat to yield, primarily through disruption of carbohydrate metabolism and diminished carbon allocation to develop. In addition to daytime heat stress, elevated night-time temperatures—commonly referred to as heat stress at night (HNT)—can also adversely affect maize productivity. During the silking phase, HNT accelerates developmental rates, which, in turn, limits the accumulation and translocation of photoassimilates to developing kernels, thereby impairing kernel set (Cantarero et al., 1999). Furthermore, HNT during the grain-filling stage has been shown to reduce kernel weight by enhancing respiratory losses and decreasing carbohydrate availability (Cantarero et al., 1999). The rise in nocturnal respiration leads to accelerated degradation of stored carbohydrates and reduces net carbon gain. HNT also disrupts the source–sink carbon dynamics by affecting the activity of key enzymes involved in non-structural carbohydrate metabolism, such as soluble starch synthase and starch branching enzymes (Yang et al., 2018; Kettler et al., 2022). These metabolic perturbations negatively influence

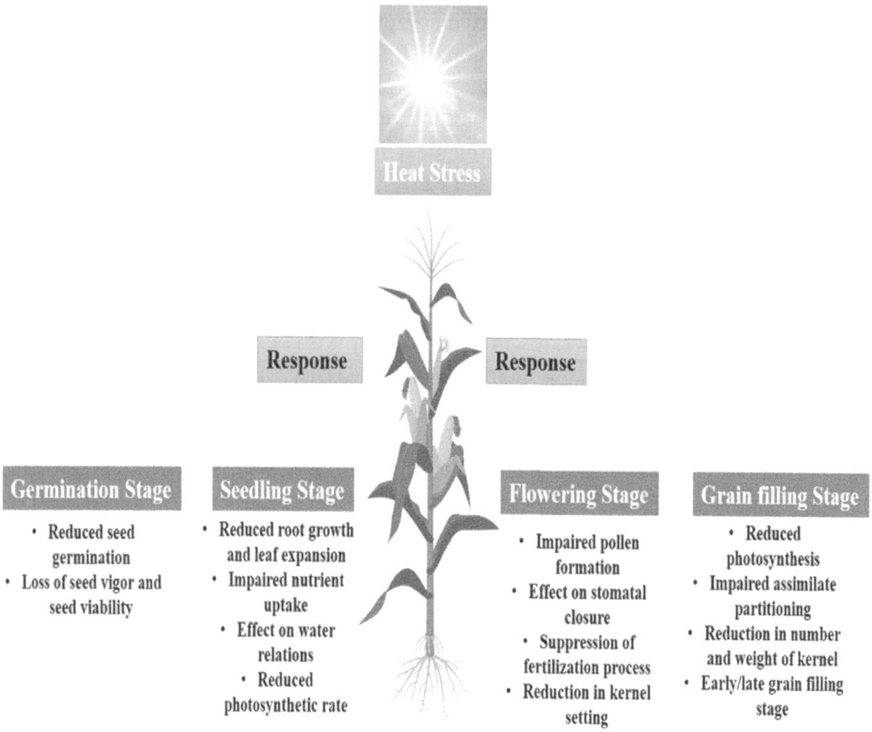

FIGURE 4.4 Heat stress and its response at various stages of maize.

assimilate partitioning and kernel development. While the adverse impacts of heat stress on maize are well documented, its intensity and outcomes are influenced by multiple interacting environmental factors. Variables such as relative humidity, vapor pressure deficit (VPD), soil moisture availability, and solar radiation significantly modulate the effects of ambient temperature on crop physiology and yield (Kebede et al., 2012; Carter et al., 2016; Thayil et al., 2020). These covariates influence crucial physiological processes, including flowering, phenological transitions, and seed setting (Bahuguna & Jagadish, 2015).

For instance, Carter et al. (2016) demonstrated that tropical maize grown under irrigated conditions exhibited minimal yield loss under high temperatures (32–34°C), whereas substantial yield reductions were observed in rainfed maize systems. Their findings suggest that soil moisture deficit, rather than heat stress alone, predominantly determines the severity of yield loss under high-temperature regimes. In addition to soil water availability, atmospheric moisture conditions, especially VPD—a function of maximum temperature and relative humidity—play a crucial role in regulating plant responses to heat stress (Thayil et al., 2020). High VPD increases transpiration demand and accelerates water loss, further exacerbating the negative impacts of heat stress under suboptimal moisture conditions. In addition to daytime heat stress, elevated night-time temperatures—commonly referred to as HNT—can also adversely affect maize productivity. During the silking phase,

HNT accelerates developmental rates, which, in turn, limits the accumulation and translocation of photoassimilates to developing kernels, thereby impairing kernel set (Cantarero et al., 1999). Furthermore, HNT during the grain-filling stage has been shown to reduce kernel weight by enhancing respiratory losses and decreasing carbohydrate availability (Cantarero et al., 1999). The rise in nocturnal respiration leads to accelerated degradation of stored carbohydrates and reduces net carbon gain. HNT also disrupts the source–sink carbon dynamics by affecting the activity of key enzymes involved in non-structural carbohydrate metabolism, such as soluble starch synthase and starch branching enzymes (Yang et al., 2018; Kettler et al., 2022). These metabolic perturbations negatively influence assimilate partitioning and kernel development.

While the adverse impacts of heat stress on maize are well documented, its intensity and outcomes are influenced by multiple interacting environmental factors. Variables such as relative humidity, VPD, soil moisture availability, and solar radiation significantly modulate the effects of ambient temperature on crop physiology and yield (Kebede et al., 2012; Carter et al., 2016; Thayil et al., 2020). These covariates influence crucial physiological processes, including flowering, phenological transitions, and seed setting (Bahuguna & Jagadish, 2015).

For instance, Carter et al. (2016) demonstrated that tropical maize grown under irrigated conditions exhibited minimal yield loss under high temperatures (32–34°C), whereas substantial yield reductions were observed in rainfed maize systems. Their findings suggest that soil moisture deficit, rather than heat stress alone, predominantly determines the severity of yield loss under high-temperature regimes. In addition to soil water availability, atmospheric moisture conditions, especially VPD—a function of maximum temperature and relative humidity—play a crucial role in regulating plant responses to heat stress (Thayil et al., 2020). High VPD increases transpiration demand and accelerates water loss, further exacerbating the negative impacts of heat stress under suboptimal moisture conditions. Low VPD in combination with high temperatures has been shown to substantially reduce maize yield, primarily due to restricted stomatal transpiration, which impairs evaporative cooling and elevates canopy temperature. Conversely, high VPD creates a more desiccating atmosphere by increasing the vapor gradient between leaf surfaces and the air, resulting in excessive water loss from plant tissues and inducing physiological drought (Seetharam et al., 2021). These opposing mechanisms demonstrate how both extremes of VPD—low and high—can negatively influence plant water balance and exacerbate the effects of heat stress. Field experiments conducted on spring maize across multiple locations in South Asia, representing humid and dry agroclimatic conditions, revealed that the impact of heat stress (>37°C) varies significantly depending on the surrounding air moisture levels. These results underscore the complex interaction between temperature and humidity, highlighting the importance of considering environmental context in evaluating crop heat stress responses.

In addition to its direct effects on plant physiology, heat stress has been increasingly recognized as a modulator of plant-pathogen interactions. Elevated temperatures can compromise plant immunity by interfering with the synthesis of key defense molecules. For example, heat stress disrupts the biosynthesis of SA—a phytohormone critical for activating systemic acquired resistance—and downregulates defense-related enzymes, thereby impairing the plant's innate immune response (Cohen & Leach, 2020). Earlier studies by Duke and Doehlert demonstrated that elevated temperatures suppress the accumulation of pathogenesis-related proteins and enzymes in maize kernels, reducing their defensive capacity against pathogenic invasion. A notable example is the increased incidence of *Aspergillus* ear rot under conditions of high temperature coupled with prolonged drought, which creates an ideal environment for fungal proliferation and toxin production (Baroševič et al., 2022). Despite growing concerns about the intersection of heat stress and pathogen susceptibility, especially under field conditions, studies investigating the combined effects of these stresses on maize in natural environments remain extremely limited. This represents a critical research gap in understanding maize resilience under future climate scenarios.

4.5.2 Physiological Responses

Exposure to heat stress detrimentally impacts a wide range of physiological and developmental processes in maize, including seed germination and vigor, root elongation, leaf expansion, reproductive development, ASI, photosynthetic efficiency, grain-filling duration, and the rate of assimilate accumulation. These impairments cumulatively reduce yield potential and compromise grain quality. At the cellular level, heat stress induces the rapid overproduction of ROS, leading to oxidative damage, cellular membrane disruption, protein degradation, and suppression of chlorophyll biosynthesis (Tas, 2022).

One of the critical physiological consequences of heat stress is its effect on stomatal conductance, which alters gaseous exchange and leads to reduced tissue water potential, impairing the plant's ability to regulate transpiration and cooling (Urban et al., 2017). Coşkun et al. (2011), in their evaluation of ten maize inbred lines under heat stress, demonstrated significant changes in physiological parameters, including increased leaf temperature (from 36.6°C to 39.2°C), elevated relative injury (RI), and decreased chlorophyll a/b content—indicators of compromised photosynthetic function. The study further reported that elevated RI corresponded with reduced membrane integrity and accelerated senescence, ultimately resulting in insufficient photoassimilate accumulation in grains. A negative correlation was observed between cellular membrane damage (CMD) and leaf water potential (LWP), indicating that as CMD increased, LWP declined—coinciding with chlorophyll loss and degradation of chlorophyll pigments. Moreover, susceptible genotypes exhibited a reduction in total phenol content (TPC) under heat stress, suggesting weakened antioxidative defense mechanisms (Tas, 2022).

Given the complexity and environmental dependence of heat stress responses, trait-based phenotyping tailored to specific environmental conditions has been recommended as an effective approach to improving thermotolerance in maize (Hall, 2011). Numerous traits associated with heat stress avoidance and tolerance have been identified. However, trait selection focused specifically on flowering and grain-filling stages is particularly crucial, as these are the most heat-sensitive developmental windows in maize (Karim et al., 1999; Commuri & Jones, 2001; Thayil et al., 2020). Targeting these phenological stages in breeding programs could significantly enhance yield stability under high-temperature stress.

Several morpho-physiological traits have been identified as potential indicators of heat stress tolerance in maize, particularly during the early vegetative or seedling stage. Among these, leaf area, leaf elongation rate, and photosynthetic efficiency have shown promise as reliable screening parameters under high-temperature conditions (Karim et al., 1999). These traits directly influence biomass accumulation and early vigor, which are critical for establishing a robust crop stand capable of withstanding subsequent stress events. Building on this, Nelimor et al. (2019) evaluated a comprehensive set of 13 traits associated directly or indirectly with GY under heat stress, classifying them into three hierarchical orders based on their predictive strength. **First-order traits**, which had the most direct impact on GY, included ear aspect, plant aspect, and root lodging—with ear aspect emerging as the most influential. **Second-order traits**, such as ears per plant, husk cover, plant height, and days to 50% anthesis, were indirectly related to GY but contributed substantially through their association with ear and plant aspects. Notably, ears per plant and husk cover showed strong correlations with both yield-determining components. **Third-order traits** included days to 50% silking, ASI, ear rot, stay-green characteristics, TSBL, and ear height. Although these traits had an indirect effect on GY, ASI and ear rot were notably influential due to their close linkage with reproductive success and ear development under heat stress.

Additionally, traits such as delayed senescence (stay-green), number of kernels per row, tassel sterility, pollen viability, and stigma receptivity have been reported as sensitive markers of thermotolerance, particularly during the reproductive phase (Alam et al., 2017; Noor et al., 2019a). These parameters serve as key physiological and reproductive benchmarks for evaluating genotypic resilience to high temperature, offering valuable insights for trait-based selection in maize breeding programs.

4.5.3 Genetic Variation for Heat Stress Tolerance

Plant responses to heat stress involve intricate physiological, biochemical, and molecular mechanisms that span multiple biological scales—from cellular components to whole-plant traits, many of which are polygenic in nature. A single episode of heat stress has been shown to trigger the differential expression of over 5,000 genes in maize (Joshi et al., 2021), underscoring the complexity of the thermotolerance response. Key adaptive traits such as pollen viability, ASI, and tassel fertility are regulated by multiple genes or QTLs, highlighting the polygenic basis of heat stress tolerance (Parrado et al., 2021).

Given this complexity, the identification and utilization of heat-responsive QTLs and genes are essential for the development of high-yielding, heat-resilient maize cultivars (McNellie et al., 2018; Inghelandt et al., 2019). The advent of advanced genomics tools, including GWAS, MAS, and high-throughput phenotyping platforms, has significantly accelerated the discovery of loci and candidate genes involved in thermotolerance. Numerous QTLs associated with heat stress tolerance have been identified (Table 4.1), and several heat-inducible genes and TFs have been comprehensively reviewed by El-Sappah et al. (2022). These molecular components offer valuable targets for precise genetic engineering and molecular breeding strategies aimed at improving heat tolerance.

However, a major limitation in current elite maize cultivars is the narrow genetic base, especially for traits associated with thermotolerance (Nelimor et al., 2019). Therefore, broadening the genetic pool through the exploitation of diverse germplasm—including landraces and wild relatives—presents a promising avenue for identifying novel alleles and adaptive traits under heat stress (Nelimor et al., 2019). Successful integration of landraces into breeding pipelines requires intermediate steps such as (i) the targeted selection of landrace(s) with desirable phenotypes, (ii) thorough evaluation for intra-population genetic variability, and (iii) development of representative core collections (Würschum et al., 2022). These foundational steps are critical to harnessing the untapped potential of native germplasm for developing next-generation heat-tolerant maize varieties.

To enhance the genetic diversity for heat stress tolerance in maize, landraces must first be top-crossed with elite cultivars to assess their combining ability and compatibility with contrasting gene pools. Depending on the diversity of traits and the variation present in the landraces, either single or multiple landraces may be selected for inclusion in crop improvement programs (Würschum et al., 2022). The goal is to develop homozygous lines from these landraces, which can then be used for association mapping studies to uncover major QTLs linked to heat stress tolerance. For example, Yuan et al. (2019) demonstrated that association mapping could identify significant QTLs responsible for the expression of quantitative traits like thermotolerance.

The efficiency of association mapping hinges on factors such as the size of the mapping population and its resolution, both of which are influenced by the linkage disequilibrium (LD) structure of the QTLs. Notably, LD decays more rapidly in landraces than in elite cultivars. Interestingly, the rate of LD decay was observed to be even faster in a diverse collection of maize genotypes, although their adaptability to adverse environmental conditions was found to be suboptimal (Würschum et al., 2022). Despite this, landraces remain a valuable resource for GWAS targeting thermotolerance traits, providing an opportunity to identify QTLs with large-effect sizes as well as the dominant genes responsible for these traits (Nelimor et al., 2019; Ahmed et al., 2022).

SNP markers are commonly employed in association studies to pinpoint genetic variations linked to heat stress tolerance (Sivakumar et al., 2019). For instance, Ahmed et al. (2022) sequenced 275 maize inbred lines grown under two different temperature regimes (35°C and 45°C) and identified a total of 170,098 SNPs. Through GWAS, they identified beneficial alleles associated with pollen germination under heat stress, revealing that each line possessed between 4–10 beneficial alleles at 35°C, and 3–13 beneficial alleles at 45°C. These alleles can be strategically utilized in breeding programs aimed at developing heat-resilient tropical maize varieties. Genes encoding heat shock factors (HSFs) play a crucial role in developing genetically modified heat-tolerant crops. In maize, 31 HSFs, including HSFA1, HSF3, HSF4, HSF5, HSF6, HSF23, and HSF25, have been identified and are known to regulate the expression of various HSPs, thereby contributing to heat stress tolerance (Lin et al., 2011). Transgenic maize expressing these HSFs has shown promising results under heat stress. For instance, overexpression of OsMYB55 resulted in higher biomass accumulation and reduced leaf damage during and after heat stress exposure, as compared to wild-type plants (Casaretto et al., 2016). Additionally, OsMYB55-overexpressing (OsMYB55-OE) plants also exhibited tolerance to drought stress (Casaretto et al., 2016).

In contrast, maize plants exposed to high night temperatures often experience GY loss due to the reduced activity of chloroplast-localized enzyme 6-phosphogluconate dehydrogenase (6PGDH or PGD3), which impairs kernel starch accumulation (Cantarero et al., 1999). While 6PGDH has two other isoforms localized in the cytoplasm (PGD1 and PGD2), only PGD3 is heat-sensitive (Spielbauer et al., 2013). Ribeiro et al. (2020) addressed this issue by fusing a chloroplast-targeting peptide sequence, Waxy1, with pgd1 and pgd2. The fusion proteins, WPGD1 and WPGD2, were able to complement the function of PGD3, leading to higher GY and biomass under high night temperature conditions.

However, despite these advancements, single-gene transgenic approaches have not been entirely successful in developing heat-tolerant maize varieties, primarily due to the complexity of the heat stress tolerance trait (Noor et al., 2019a; Jha et al., 2021). Recently, genome editing techniques such as ZFN, TALEN, and CRISPR-Cas have been employed to develop abiotic stress-tolerant crops (Ainley et al., 2013; Char et al., 2015; Bhat et al., 2021). For example, CRISPR-Cas9 was used to enhance drought tolerance in maize by targeting the maize GSO2 promoter and inserting it into the 5′-untranslated region (5′-UTR) of the ARGOS8 gene, resulting in improved drought tolerance (Shi et al., 2017). Similarly, Zhu et al. (2016) utilized type-II CRISPR-Cas9 to target the maize phytoene synthase gene (PSY1) and investigate its role under heat stress. They found that PSY1 plays a key role in thermotolerance by maintaining carotenoid synthesis in leaves, which helps mitigate oxidative damage under heat stress.

4.5.4 Molecular Mechanisms

Plants respond to environmental stress, such as heat, by repairing damage to cellular components and enhancing their defense mechanisms for future stress exposures (Li & Howell, 2021). Heat stress responses at the cellular level can be categorized into two major types: (i) the heat shock response, which primarily occurs in the cytoplasm, and (ii) the unfolded protein response (UPR), which is generated in the endoplasmic reticulum (ER; El-Sappah et al., 2022). These responses are triggered by the accumulation of misfolded proteins and are essential for maintaining cellular integrity under stress conditions. In addition to these responses, heat stress also initiates specific responses at the plasma membrane, beginning with changes in membrane fluidity. These changes can open voltage-gated Ca^{2+} channels and trigger ROS production in the chloroplast thylakoid membrane (, 2012; Finka et al., 2012; Bahuguna & Jagadish, 2015;). An increase in membrane fluidity and the subsequent efflux of Ca^{2+} into the cytoplasm are considered primary responses to heat stress (Finka et al., 2012; Mittler et al., 2012). Calcium signals activate MAPKs, which, in turn, activate TFs such as HSFA6b, ABF1, CYD5, and MutS2, all of which are specific to heat stress tolerance in maize (Gao et al., 2012).

Heat stress can also be sensed by plasma membrane-bound receptor proteins. These receptors include cyclic nucleotide-gated channels (CNGCs) and glutamate receptor-like channels, which are involved in heat stress tolerance (Liu et al., 2018). Several Ca^{2+}-permeable channels have been identified in plants, including voltage-dependent, voltage-independent, depolarization-activated, and hyperpolarization-activated Ca^{2+}-permeable channels (Swarbreck et al., 2013). In maize, 11 CNGCs have been reported (Hao & Qiao, 2018), each with two binding domains: one for cyclic nucleotides and one for calmodulin (CaM) in the cytoplasmic region (Zelman et al., 2012). When cyclic nucleotides bind to the cyclic nucleotide binding domain, they trigger the influx of Ca^{2+}, which then binds to CaM. The Ca^{2+}–CaM complex activates various TFs such as ZmCAMTA, ZmbZIP, ZmMYB, and ZmWRKY-R. Notably, ZmCAMTA 1, 2, and 3 are highly expressed under heat stress conditions (Yue et al., 2015; El-Sappah et al., 2022). These TFs bind to heat shock elements (HSEs) in the promoter regions of HSP genes, thereby inducing their expression (Zhang et al., 2020b).

Additionally, elongation factors such as EF-Tu, known for their role in protein synthesis, have been associated with heat stress tolerance in crops, including maize (Momčilović & Ristic, 2007; Ristic et al., 2008; Fu et al., 2012). Beyond the CaM-mediated pathway, CNGCs and glutamate heat receptor-like channels transfer stress signals through the activation of Ca^{2+}-dependent protein kinases, calcineurin B-like proteins, and interacting protein kinases (CIPK), along with phosphoinositide-specific phospholipase C (Reddy et al., 2004; Chen et al., 2011; Hashimoto & Kudla, 2011; Zhao et al., 2021). These enzymes hydrolyze membrane-bound phosphatidylinositol-4,5-bisphosphate (PIP_2) into inositol trisphosphate (IP_3) and diacylglycerol, which, upon activation by G-protein coupled receptors (GPCRs), trigger the release of Ca^{2+} from the ER, thus continuing the Ca^{2+}-mediated signaling cascade (Ruelland et al., 2002; Horváth et al., 2012). Phosphoinositide-specific phospholipase C (PLC) isoforms such as PLC3, PLC9, and ZmPLC1 (in maize) have been particularly linked to heat stress tolerance (Rupwate & Rajasekharan, 2012; Zhai et al., 2013; Hayes et al., 2021). At the end of this signaling cascade, activated CDPKs phosphorylate heat shock factors (HSFs), which are then imported into the nucleus to promote the expression of HSPs such as HSP40, HSP70, HSP90, and HSP100 (Qian et al., 2019).

Heat shock factors (HSFs) are categorized into three classes: A, B, and C. Among the 31 HSFs identified in maize, some key members associated with heat stress tolerance include ZmHsf1, 4, 5, 6, and 17 (class A), as well as ZmHsf3, 11, and 25 (class B) (Lin et al., 2011; Zhang et al., 2020a). Notably, only class A HSFs contain a transcriptional activation domain (Kotak et al., 2004). HSFA1, in particular, is considered the master transcription activator in response to heat stress. Under normal conditions, HSFA1 is sequestered in the cytoplasm by binding to chaperones (HSP70 and HSP90) and co-chaperones (Ohama et al., 2017). However, during heat stress, the chaperones are released from HSFA1, allowing it to trimerize and be transported into the nucleus, where it activates the expression of heat stress-responsive genes (Yan et al., 2020; Zhao et al., 2021). A parallel signaling cascade is triggered by the production of ROS under heat stress (Hasanuzzaman et al., 2012, 2013). Membrane-bound NADPH oxidase, which is part of the respiratory burst oxidase homolog D family, produces superoxide radicals (O_2^-) from molecular oxygen (O_2). The half-life of O_2^- is very short, and it is rapidly dismutated into a more stable ROS species, hydrogen peroxide (H_2O_2), either spontaneously or by the apoplastic SOD enzyme

(Bahuguna & Jagadish, 2015). H_2O_2 then enters the cytoplasm, where it activates several HSFs, either directly or indirectly. In maize, H_2O_2 activates MAPKs, specifically MAPK3 and MAPK6, which, in turn, activate CDPKs. This cascade of events ultimately leads to the phosphorylation of HSFs and the induction of HSP expression (Luna et al., 2011; Matika & Loake, 2014).

In addition to its role in the MAPK pathway, H_2O_2 directly activates HSFA1α, HSFA4α, and HSFA8, along with nitric oxide (NO) signaling, to further induce the expression of heat response genes in maize (El-Sappah et al., 2022). Furthermore, in addition to the signaling initiated at the plasma membrane, a specific UPR occurs in the ER when the concentration of misfolded proteins increases rapidly under heat stress (Vitale & Boston, 2008; Neill et al., 2019). Misfolded proteins recruit the binding immunoglobulin protein (BIP), which dissociates from the noncompetent, monomeric form of inositol-requiring enzyme-1 (IRE1) (Carrara et al., 2015). BIPs assist misfolded proteins in binding to IRE1, leading to its dimerization. The dimeric form of IRE1 functions as both a ribonuclease and a kinase, which facilitates the splicing of basic leucine zipper domain (ZmbZIP60) mRNA. This splicing allows the mRNA to encode the ZmbZIP60 TF, which is specific to heat stress and is imported into the nucleus to initiate heat stress response (Reimold et al., 2000). Another mechanism that activates the UPR in the ER involves the proteolytic cleavage of ER-membrane proteins, ZmbZIP17 and ZmbZIP28, in the Golgi apparatus. The cleaved globular domains of these proteins act as TFs and are transported to the nucleus, where they activate the expression of HSPs (Jia et al., 2009; Zhang et al., 2017). In addition to ROS and calcium (Ca^{2+}), other secondary messengers, including phytohormones such as IAA, CK, GA, ABA, SA, ET, BRs, and jasmonic acid, have also been implicated in the signal transduction pathways that mediate heat stress tolerance in plants (Eyidogan et al., 2012; Li et al., 2021). For example, Huang et al. (2016) demonstrated that the application of ABA, along with the expression of the AREB1/ABF1 TF, enhanced thermotolerance in maize as shown in Figure 4.3.

While the heat stress response in maize is relatively well-characterized, it is important to note that maize plants exposed to prolonged heat stress can experience physiological drought stress, especially under rainfed conditions (Seetharam et al., 2021). Furthermore, combined heat and drought stress is a growing concern, particularly in hot and dry environments, and is expected to increase with changing climates (Haddad et al., 2022). Despite the significance of this combination, there are limited studies that focus exclusively on genes expressed under the combined heat and drought stress conditions. Qin et al. (2007) found that maize plants overexpressing ZmDREB2A (ZmDREBA2-OE) exhibited enhanced drought tolerance by activating the expression of late embryonic abundant genes. This TF also promoted the upregulation of genes related to detoxification and thermotolerance, with four specific genes, including At5G03720, being upregulated by sevenfold under heat stress. Similarly, the OsERF115/AP2EREBP110 gene in rice, which belongs to the AP2/EREBP TF family, plays a crucial role in both heat and drought tolerance (Suzuki et al., 2016). Thus, identifying genes that are specifically expressed under the combined heat and drought stress conditions is critical for understanding maize's response to complex environmental stress interactions across spatial and temporal scales.

4.6 HEAVY METALS STRESS: EFFECTS ON GROWTH AND YIELD

Heavy metal toxicity poses a significant threat to plant health, disrupting physiological processes and leading to substantial yield losses. Moreover, these toxic elements can accumulate in plant tissues and enter the food chain, ultimately affecting human and animal health. To mitigate these impacts, plants have evolved complex molecular and cellular mechanisms that enhance their tolerance and reduce heavy metal accumulation. Understanding and harnessing these adaptive responses is crucial for improving both crop safety and environmental health.

Plant species differ widely in how they respond to heavy metal stress—ranging from sensitivity to tolerance, and in some cases, even hyperaccumulation. Currently, two major biotechnological strategies are being pursued to address heavy metal contamination. The first focuses on increasing the uptake and accumulation of metals in specific plant species for use in phytoremediation and phytoextraction—techniques that clean up contaminated soils. The second strategy aims to enhance plant resistance to heavy metal toxicity, thereby safeguarding growth and productivity (Ovečka & Takáč, 2014). Beyond genetic traits, several environmental factors influence the severity and effects of heavy metal toxicity. Soil characteristics (edaphic factors), light conditions, the chemical nature and concentration of metals, and their oxidation states all play key roles in determining how metals are absorbed, transported, and distributed within plant tissues (Emamverdian et al., 2015). Regardless of their tolerance level, all plants possess a basic defense system that activates upon sensing heavy metal stress. Recent advancements in molecular biology have significantly deepened our understanding of these intricate defense mechanisms, shedding light on how plants perceive, respond to, and adapt to toxic metal exposure.

4.6.1 Heavy Metal Stress in Plants

Plants in their natural environments are constantly exposed to a range of abiotic and biotic stresses. Among these, heavy metal contamination has become an escalating concern due to the persistent accumulation of elements such as iron (Fe), manganese (Mn), copper (Cu), nickel (Ni), cobalt (Co), cadmium (Cd), zinc (Zn), mercury (Hg), and arsenic in soils. This buildup is largely driven by anthropogenic activities, including industrial discharge, excessive fertilizer use, smelting operations, and improper sewage disposal (Aydinalp & Marinova, 2009). These processes either lead to metal leaching into groundwater or their accumulation on the soil surface (Gupta & Ali, 2002; Ali et al., 2009; Aydinalp & Marinova, 2009; Gupta & Ali, 2012; Dağhan & Ozturk, 2015; Basheer, 2018a).

Unlike organic pollutants, heavy metals are non-biodegradable and cannot be broken down by natural processes. While some metals remain immobile and tend to accumulate locally in soils, others are mobile and can be absorbed by plant roots. This uptake occurs through various mechanisms such as passive diffusion, endocytosis, or active transport via specific metal transporters (Ali & Jain, 2004; Ashraf et al., 2010a; Krzesłowska, 2011; Sabir et al., 2015; Dehghani et al., 2016; Ali et al., 2018; Alharbi et al., 2018; Burakova et al., 2018).

Interestingly, not all heavy metals are harmful. Trace elements like zinc, copper, and nickel are essential micronutrients that serve as cofactors in enzymatic reactions and play vital roles in plant metabolism. However, other metals such as cadmium and lead—often introduced through pesticides and industrial pollutants—have no known biological function and become toxic when their concentrations surpass threshold levels (Genchi et al., 2020; Ali et al., 2011, 2017b). Plants naturally act as bioaccumulators, meaning they have the ability to absorb and concentrate various heavy metals from soil and water, regardless of whether these elements are beneficial or harmful to their growth. This characteristic has both ecological risks and potential applications in environmental remediation (Aksoy et al., 2000; Ali & Aboul-Enein, 2002; Yilmaz et al., 2006; Gücel et al., 2009b; Celik et al., 2010; Haribabu & Sudha, 2011; Ali et al., 2016b; Ozturk et al., 2017).

The capacity of plants to tolerate and accumulate heavy metals differs significantly among species. Some plants are more sensitive and exhibit toxicity symptoms more rapidly than others. Common signs of heavy metal stress in plants include stunted growth, yellowing of leaves (chlorosis), browning of roots, progressive decline, and ultimately plant death (Ozturk et al., 2008, 2015b).

At the biochemical level, heavy metals disrupt vital physiological processes by inactivating or denaturing key enzymes and proteins. They also displace essential metal ions from their binding sites on biomolecules, leading to structural and functional impairments. These disruptions compromise membrane integrity and interfere with fundamental metabolic activities such as photosynthesis, respiration, and nutrient homeostasis (Hossain et al., 2012). Additionally, heavy metal stress triggers the overproduction of ROS, including superoxide radicals (O_2^-), hydroxyl radicals (•OH), and hydrogen peroxide (H_2O_2). These ROS cause oxidative damage to cellular components, exacerbating stress and further impairing plant function. In addition to ROS, plants under heavy metal stress also show elevated levels of another cytotoxic compound called methylglyoxal. This highly reactive molecule contributes significantly to oxidative stress by inducing lipid peroxidation—particularly in cellular membranes—leading to membrane leakage, damage to vital biomolecules, and even DNA strand breaks (Romero-Puertas et al., 2002; Hossain et al., 2010; Barconi et al., 2011; Ahmad et al., 2012).

To counter these harmful effects, plants have evolved a range of molecular, physiological, and biochemical defense mechanisms. These include both constitutive (always active) processes that are part of the plant's basic homeostatic system, and inducible responses that are activated only upon exposure to specific heavy metals. These mechanisms operate at multiple levels—from metal sequestration and detoxification to antioxidant defense and gene regulation. Broadly, plant responses to heavy metal stress can be categorized into two main strategies: avoidance, which minimizes metal uptake or limits its mobility within the plant; and tolerance, which enables the plant to endure and function despite the presence of toxic metal concentrations (Krzesłowska, 2011).

When exposed to heavy metal stress, plants initiate a multilayered defense strategy to mitigate toxicity. The first line of defense involves minimizing metal uptake at the root level. This is achieved through cellular processes and the release of root exudates, which either bind metals to the cell wall (biosorption) or promote their efflux, preventing entry into the cell. These mechanisms are considered part of the plant's "avoidance" strategy. In addition to avoidance, many plants employ "tolerance" mechanisms that deal directly with internalized metal ions. These include sequestration of metals into specific cellular compartments—such as vacuoles—to isolate them from sensitive cellular machinery. Protective compounds like amino acids, glutathione, phytochelatins, and metallothioneins, along with detoxifying enzymes such as SOD and POD, play key roles in these processes (Hossain et al., 2012).

However, under prolonged or extreme metal stress, these routine defense systems may become overwhelmed. In such cases, plants activate more specialized detoxification pathways. These involve chelation, transport, compartmentalization, or chemical transformation of metals to less toxic forms, often within the vacuole. The stress response also includes the synthesis of specific proteins and signaling molecules such as stress hormones, antioxidants, and HSPs to support cellular recovery and survival. Plants also adopt ecological strategies to manage heavy metal stress. For example, forming symbiotic relationships with mycorrhizal fungi can help immobilize metals in the rhizosphere, limiting their uptake by roots. At the molecular level, plants rely on complex and highly regulated signaling networks to coordinate these responses. One of the most crucial pathways in abiotic stress signaling is the MAPK cascade, which is evolutionarily conserved among eukaryotes (Tena et al., 2001; Jonak et al., 2002). This pathway, driven by a series of phosphorylation events, regulates diverse cellular outcomes, including gene expression, cell division, differentiation, and cross-talk with hormonal signals. Notably, the impact of heavy metals on plant growth can be either inhibitory or stimulatory depending on their concentration and speciation in the soil (Abolghassem et al., 2018). Most heavy metals tend to accumulate in root tissues, where barriers such as the Casparian strip and cell wall binding restrict their movement into the shoot. Despite these defenses, metal accumulation can impair physiological, biochemical, and morphological functions, ultimately reducing crop productivity (Shahid et al., 2015).

In conclusion, plants constantly face abiotic stresses such as heavy metal toxicity and have developed an array of dynamic and evolving strategies to cope with them. This review highlights the toxic effects of various heavy metals and provides an overview of the plant responses that enable survival and adaptation under such challenging conditions.

4.6.2 Impact of Heavy Metal Stress in Maize

The problem of heavy metal pollution in the agricultural field is aggravated due to their persistent nature in the environment. Heavy metal stress poses a significant threat to maize growth, yield, and grain quality. Multiple studies have demonstrated that exposure to metals such as cadmium (Cd), mercury (Hg), lead (Pb), chromium (Cr), and copper (Cu) leads to oxidative stress, disruption of mineral uptake, alterations in root and shoot morphology, and inhibition of physiological and biochemical processes.

In one study, Cd and Hg exposure caused growth inhibition and oxidative damage in maize seedlings, with Hg proving more toxic. Activation of cytosolic APX and phytochelatin accumulation in Cd-treated plants suggested a partial defense mechanism, while Hg exposure resulted in greater oxidative imbalance and reduced glutathione content. (Rellán-Álvarez et al., 2006). Similarly, Pb and Cr at higher concentrations (300 μM) caused significant reductions in photosynthetic pigments, protein content, and mineral uptake, accompanied by lipid peroxidation and hydrogen peroxide accumulation. Notably, Cr exhibited stronger toxicity even at lower concentrations, with marked anatomical alterations, including degradation of cortical cells and thickening of the endodermis and leaf midrib—indicative of structural adaptations to limit metal translocation and maintain physiological function (Singh et al., 2015). Further molecular-level insights revealed that the high-affinity sulfate transporter gene **ZmST1;1** in maize roots is transcriptionally regulated under Cd stress, linking sulfate acquisition to thiol metabolism and phytochelatin biosynthesis. This suggests that enhanced sulfur uptake is a key adaptive response to heavy metal exposure (Nocito et al., 2006). In addition to inherent defense mechanisms, external applications of silicon (Si) and silicon nanoparticles (SiNPs) demonstrated effective mitigation of Cd-induced stress. Foliar spraying with Si and SiNPs improved plant growth and antioxidant enzyme activities while reducing oxidative stress markers and Cd accumulation in grains. Although both forms of silicon showed similar protective effects, their role in enhancing stress resilience and reducing metal toxicity is promising for sustainable crop management (Rahman et al., 2023). Biological approaches, such as microbial inoculation with metal-tolerant rhizobacteria, also offer practical remediation strategies. *Azotobacter chroococcum* strain CAZ3, isolated from metal-polluted rhizospheres, enhanced maize growth, kernel quality, and reduced oxidative damage under Cu and Pb stress. The strain maintained plant growth-promoting traits under metal exposure and sequestered metals primarily in roots, minimizing translocation to edible tissues (Rizvi & Khan, 2018). Finally, studies focusing on copper stress revealed extensive physiological and morphological damage to maize roots, including decreased biomass and disrupted mineral content. However, ultrastructural analysis showed variability in cellular damage, with some cells maintaining intact organelles and membrane structures. This heterogeneity in cellular response indicates a potential for localized resistance strategies in maize roots against Cu toxicity (Ouzounidou et al., 1995).

In summary, maize exhibits a range of physiological, molecular, anatomical, and biochemical responses to cope with heavy metal stress (Table 4.5).

Understanding these responses not only deepens our knowledge of plant stress physiology but also supports the development of integrated strategies—including molecular regulation, mineral supplementation, and microbial inoculants—for improving maize resilience in contaminated environment.

TABLE 4.5
Responses of Maize to Heavy Metal Stress

Heavy Metal	Observed Effects	Physiological/Biochemical/ Molecular Response	Reference
Cadmium (Cd)	Growth inhibition, oxidative damage	Increased cytosolic APX activity, phytochelatin accumulation, altered sulfate transporter (ZmST1;1) expression	Rellán-Álvarez et al. (2006); Nocito et al. (2006)
Mercury (Hg)	Greater oxidative stress, reduced glutathione content	High toxicity compared to Cd; severe oxidative imbalance	Rellán-Álvarez et al. (2006)
Lead (Pb)	Reduced photosynthetic pigments and protein content; lipid peroxidation	Disruption of mineral uptake; oxidative stress; sequestered mainly in roots with bacterial inoculation	Singh et al. (2015); Rizvi and Khan (2018)

(Continued)

TABLE 4.5 *(Continued)*
Responses of Maize to Heavy Metal Stress

Heavy Metal	Observed Effects	Physiological/Biochemical/ Molecular Response	Reference
Chromium (Cr)	Strong toxicity even at low concentrations; anatomical damage	Degradation of cortical cells, thickened endodermis and leaf midrib—structural adaptation to restrict movement	Singh et al. (2015)
Copper (Cu)	Reduced biomass, mineral imbalance, root damage	Cell-to-cell variability in ultrastructural damage, indicating possible localized resistance	Ouzounidou et al. (1995)
Cd (mitigation)	Cd toxicity mitigated by Si and SiNPs	Enhanced growth, increased antioxidant enzyme activities, reduced Cd accumulation in grains	Rahman et al. (2023)
Pb & Cu (mitigation)	Improved tolerance with Azotobacter chroococcum CAZ3 inoculation	Enhanced kernel quality, reduced oxidative damage, metal sequestration in roots	Rizvi and Khan (2018)

4.7 IMPACTS OF CLIMATE CHANGE AND ITS MITIGATION STRATEGIES

Over the next half-century, global agriculture faces the monumental task of feeding an additional 3.5 billion people (Borlaug, 2007). To meet this growing demand, the production of the world's three primary cereal crops—maize, wheat, and rice—must increase by approximately 70% by the year 2050. This is a critical goal for ensuring food security across both rural and urban populations worldwide. However, projections under various climate change scenarios suggest that agricultural productivity will face significant setbacks, particularly in regions already vulnerable to food insecurity (Lobell et al., 2008). Climate change, driven by the accumulation of greenhouse gases such as carbon monoxide (CO), methane (CH_4), chlorofluorocarbons (CFCs), nitrous oxide (NO), and ozone (O_3) in the atmosphere, has led to a measurable rise in global temperatures (IPCC, 2007). This warming trend is not only reshaping weather patterns but also amplifying the frequency and intensity of extreme weather events such as droughts, floods, and heatwaves—phenomena that severely disrupt agricultural systems. In developing countries, where adaptive capacity is often limited, climate variability remains the leading cause of fluctuations in food production. Many regions experience low mean annual rainfall coupled with high temporal variability, compounding the uncertainty of crop yields. The effects of climate change on crop production are complex and highly localized, varying across regions, seasons, and crop species. Key climatic variables such as temperature, precipitation, soil moisture, and atmospheric pressure directly influence plant growth, either individually or through interactions with other environmental factors.

This climate-induced instability poses a dual threat to rural livelihoods and global food security. It undermines not only the socioeconomic foundations of agricultural communities (Burke et al., 2009) but also the ecological systems that support them (Walker & Schulze, 2008). These challenges are especially severe in low-income, resource-constrained nations where resilience to climate stress is minimal, making them particularly susceptible to the cascading effects of food shortages, economic disruption, and ecological degradation. Despite extensive research on the broader relationship between climate and agriculture, comparatively few studies have focused specifically on the impact of climate change on maize production. Maize (*Zea mays* L.), a member of the Poaceae family, is the third most significant staple crop globally, following wheat and rice. It plays a central role not only in human and animal nutrition but also in a wide range of industrial applications. As a multipurpose crop, maize serves as a critical raw material in the production of starch, oil, proteins, alcoholic beverages, biofuels, sweeteners, pharmaceuticals, cosmetics, biodegradable films, textiles, adhesives, and paper products. Maize is often referred to as the "Miracle Crop" due to its exceptional adaptability and photoperiod insensitivity, allowing it to thrive in diverse agroecological conditions. It is a C_4 plant, which means it possesses a more efficient photosynthetic pathway compared to C_3 crops, resulting in superior biomass accumulation and GY potential. This physiological advantage has earned maize the title "Queen of Cereals." The nutritional composition of maize further highlights its value: maize kernels contain approximately 77% starch, 2% sugars, 9% protein, 2% ash (on a dry weight basis), 5% pentosans, and 5% oil. Maize oil is particularly noteworthy for its high polyunsaturated fatty acid (PUFA) content, especially linoleic acid (61.9%), which contributes to cardiovascular health by maintaining fluidity at LTs. It also contains minimal linolenic acid (0.7%) and retains a rich, natural flavor, making it a preferred oil in culinary and nutritional applications. Globally, maize serves primarily as a food source, although in countries like the United States, up to 85% of maize production is allocated to livestock feed. However, maize production is increasingly challenged by climate variability. Shifts in weather patterns—such as erratic rainfall, prolonged droughts, or unseasonal heavy rainfall—can severely affect critical stages of maize development, including germination, flowering, and grain filling. These disruptions can result in dramatic reductions in yield and even complete crop failures. Recent years have witnessed considerable fluctuations in maize yields, largely attributable to extreme climate events. In many regions, the uncertainty of rainfall patterns has forced farmers to reduce the area under maize cultivation as a risk mitigation strategy. The steady decline in maize production underscores the urgent need for targeted research and climate-resilient agricultural strategies to sustain and enhance maize productivity in the face of mounting climatic pressures.

4.7.1 Abiotic Stresses of Maize under the Changing Climate

a. **Drought:** Drought stands as the most widespread and significant constraint limiting maize yield potential globally (Edmeades et al., 1993). Empirical estimates suggest that average annual yield losses due to drought range from approximately 15% in temperate regions to 17% in tropical zones. However, accurately quantifying global yield losses is inherently difficult, given the highly variable nature of drought events. These range in scale from localized field-level occurrences to widespread regional phenomena and vary in severity from mild water stress to devastating crop failure. The impact of drought is especially pronounced in regions with inherently poor soils and erratic climatic conditions, where agricultural systems lack the buffering capacity of more favorable zones. In contrast, areas such as the U.S. Corn Belt benefit from deep, fertile soils and relatively predictable rainfall patterns, which significantly reduce the risk of severe drought-induced yield losses. These advantages have positioned the Corn Belt as one of the most stable and productive maize-growing regions in the world. Nonetheless, even in such favorable zones, the increasing frequency and intensity of drought events driven by climate change pose a growing threat to sustainable maize production.

b. **Heat:** By the end of the 21st century, average temperatures during the growing season are projected to surpass even the most extreme seasonal highs recorded over the past hundred years (Battisti & Naylor, 2009). Historical data from crop production and meteorological records highlight the sensitivity of maize to heat stress. For instance, Thomson (1966) reported that a 6°C temperature increase during the grain-filling stage led to a 10% reduction in maize yields in the U.S. Corn Belt. More recent studies have confirmed this vulnerability; yields have been shown to decline significantly with increased exposure to daily maximum temperatures exceeding 32°C during grain filling. In sub-Saharan Africa, the impact of temperature is even more pronounced. Lobell and Burke (2010) estimated that a 2°C rise in temperature would lead to greater maize yield reductions than even a 20% decline in precipitation—underscoring heat stress as a more critical limiting factor than drought in certain contexts. Supporting this, Lobell et al. (2011) analyzed over 20,000 maize trial records from Africa over an eight-year period and found that for each additional degree day above 30°C, GYs declined by 1% under optimal rainfed conditions and by 1.7% under drought stress. Maize is particularly vulnerable during its reproductive phase, when the threshold for heat damage is notably lower in reproductive tissues than in vegetative parts (Stone, 2001). Heat stress during this stage disrupts several key reproductive processes, including the production of viable pollen, successful pollination, and proper development of the embryo and endosperm (Schoper et al., 1987a). Elevated temperatures during flowering and grain set reduce both the number of fertilized ovules and kernel weight, ultimately resulting in significant yield losses.

c. **Water Logging:** Flooding and waterlogging are significant abiotic stress factors that constrain maize production, particularly in South and Southeast Asia, where over 18% of the total maize-growing area is regularly impacted, resulting in annual yield losses ranging from 25% to 30% (Zaidi & Singh, 2010). While less extensive in sub-Saharan Africa, waterlogging remains a concern in select regions, especially those with poor drainage or irregular rainfall patterns.

Waterlogging stress arises when soil moisture exceeds the field capacity, raising the water table to levels that impede normal plant physiological processes. One of the primary consequences is the drastic reduction in the diffusion rate of gases within the soil—up to 100 times slower than in air—severely limiting oxygen exchange between root tissues and the atmosphere (Armstrong & Drew, 2002). This leads to hypoxic conditions in the rhizosphere within hours, and if the saturation persists beyond three days, the roots may experience anoxia, or complete oxygen deprivation (Zaidi & Singh, 2010). Compounding the lack of oxygen, waterlogged soils also accumulate harmful gases such as carbon dioxide, ET, hydrogen sulfide, methane, and ammonia, which are toxic to plant roots and further hinder growth. In addition, prolonged waterlogging disrupts the availability and uptake of essential nutrients. Oxygen scarcity impairs ATP production, diminishing the energy required for active nutrient transport across root membranes. Moreover, water-saturated soils alter redox potentials, leading to nutrient imbalances: nitrogen is converted from nitrate to ammonium, sulfur to hydrogen sulfide—both of which are less available to non-wetland species like maize. Iron and manganese often accumulate to toxic levels, while phosphorus availability may fluctuate depending on soil pH dynamics during saturation events.

Altogether, waterlogging exerts a multifaceted impact on maize physiology—restricting gas exchange, altering hormonal balance, and severely compromising nutrient acquisition—thereby posing a major threat to yield stability in flood-prone regions.

4.7.2 Regional Impacts and Adaptive Strategies for Maize Production under Climate CHANGE-Driven Intensification of Abiotic Stresses

Global climate variability is intensifying the frequency and severity of drought and heat stress across agricultural regions, with semiarid zones being especially vulnerable. Among staple crops, maize is particularly susceptible to these abiotic stresses, which have emerged as significant constraints on crop development and yield (Ahmad et al., 2024). Heat stress, in particular, results in reduced yields, lower grain quality, and increased susceptibility to pests and diseases. Maize responds to heat stress through a range of physiological mechanisms, including adjustments in photosynthesis and respiration, shifts in water use efficiency, and alterations in reproductive processes. These adaptations are critical for sustaining productivity under increasingly extreme climatic conditions (Figure 4.5).

Predictive models that integrate future climate scenarios and developmental responses—such as stomatal behavior—offer valuable tools for informed decision-making in crop management (Ahmad et al., 2024).

To mitigate the adverse effects of heat stress, both conventional breeding and biotechnological strategies have been employed. Genetic engineering efforts have focused on improving heat tolerance and stomatal regulation, with specific genes such as *AOX*, *Zm-AN13*, and *ZmSEC14p* identified as key contributors to thermal resilience (Ahmad et al., 2024). Combining molecular approaches with traditional breeding and physiological insights provides a robust pathway to enhancing maize resilience in the face of climate change. In Sub-Saharan Africa (SSA), climate change has already imposed significant challenges to maize-based food systems.

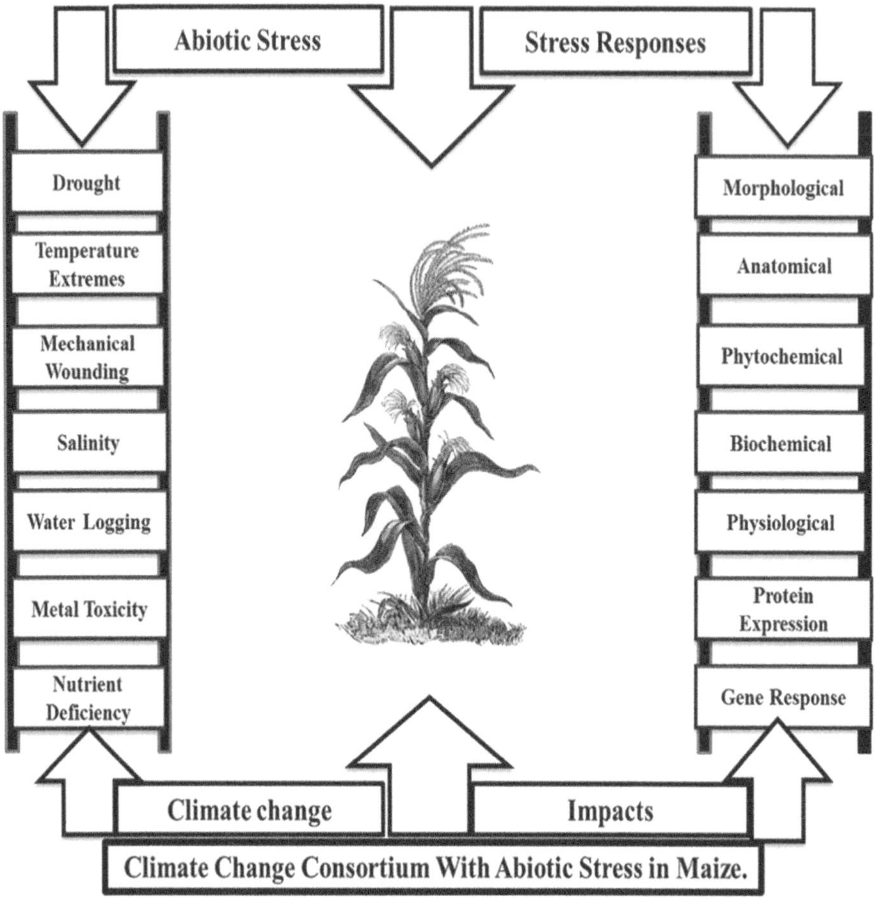

FIGURE 4.5 Climate change consortium and abiotic stress in maize.

Altered rainfall patterns, rising temperatures, and poor soil and water management have led to reduced soil moisture and severe yield declines—ranging from 30% to 90% in some areas (Zizinga et al., 2022). Projections for the region indicate continued increases in temperature and rainfall variability, further threatening agricultural productivity (Adhikari et al., 2015).

Although drought-tolerant maize varieties have been introduced, their adoption and effectiveness remain limited. Climate Smart Agriculture (CSA) practices—including mulching, halfmoon pits, and permanent planting basins—have demonstrated substantial promise in enhancing soil moisture retention, improving water use efficiency, and boosting maize yields across various agroecological zones (Sawadogo, 2011; Bu et al., 2013). However, in Uganda, uptake remains low due to limited research on optimal CSA design and suitability across different socioeconomic and climatic contexts (Zizinga et al., 2022). Tools like AquaCrop have been utilized to simulate the performance of these practices, yet further refinement is required for their effective deployment in sub-humid tropical zones (Kikoyo & Nobert, 2016; Mibulo & Kiggundu, 2018). In China, maize production is similarly affected by climate shifts, particularly in temperature, rainfall, and solar radiation. An analysis of data from 3,050 smallholder farmers between 2009 and 2018 revealed that while temperature and rainfall generally enhance maize yields, limited sunshine reduces productivity (Zhang et al., 2022). These impacts vary across the five main maize-growing regions, necessitating region-specific strategies. Enhancing weather forecasting, optimizing crop layouts, educating farmers on climate adaptation, and improving irrigation systems are essential for maintaining productivity under changing climatic conditions. In Northern Serbia, maize—the region's primary spring crop—is expected to become increasingly vulnerable in the near term (2001–2030). AquaCrop simulations indicate that projected declines in rainfed maize yields will largely result from reduced growing-season precipitation and a rise in extreme heat events, such as tropical days during summer (Daničić et al., 2021). Although agriculture in Serbia is mostly rainfed, these stressors suggest that irrigation will likely become essential not only for maize but also for other spring crops like soybean and sugar beet. Collectively, these regional perspectives underscore the urgent need for climate-resilient maize production strategies. Integrating physiological understanding, genetic innovations, CSA practices, and region-specific planning is crucial for sustaining maize yields and supporting food security amid the ongoing climate crisis.

4.8 CONCLUSION

In conclusion, the impacts of climate change, including drought, heat, and nutrient deficiencies, are severely affecting maize production worldwide. As temperatures rise and rainfall patterns shift, maize—one of the most climate-sensitive crops—faces yield reductions due to disrupted growth, especially during pollination and grain set. Studies show that for every degree Celsius increase in temperature, global maize production decreases by 8.3%, with projected yield losses of 10–20% by the century's end. This decline threatens food security, particularly in regions like Sub-Saharan Africa and the Indo-Gangetic plains, where climate variability accounts for up to 50% of yield fluctuations. With the global population expected to reach 9 billion by 2050, maize will need to adapt to meet growing food demands. To address these challenges, strategies such as developing climate-resilient maize varieties, improving water management, and enhancing climate forecasting will be crucial to sustaining maize production and ensuring food security.

REFERENCES

Abolghassem, E., Ding, Y., Farzad, M., & Xie, Y. (2018). Antioxidant response of bamboo (*Indocalamus latifolius*) as affected by heavy metal stress. *Journal of Elementology*, 23(1), 341–352.

Adhikari, U., Nejadhashemi, A. P., & Woznicki, S. A. (2015). Climate change and eastern Africa: A review of impact on major crops. *Food and Energy Security*, 4(2), 110–132.

Agami, R. A. (2013). Alleviating the adverse effects of NaCl stress in maize seedlings by pretreating seeds with salicylic acid and 24-epibrassinolide. *South African Journal of Botany*, 88, 171–177.

Ahmad, A., Tola, E., Alshahrani, T. S., & Seleiman, M. F. (2023). Enhancement of morphological and physiological performance of *Zea mays* L. under saline stress using ZnO nanoparticles and 24-eibrassinolide seed priming. *Agronomy*, 13, 771. doi: 10.3390/agronomy13030771

Ahmad, P., Ozturk, M., & Gucel, S. (2012). Oxidative damage and antioxidants induced by heavy metal stress in two cultivars of mustard (Brassica juncea L.) plants. *Fresenius Environmental Bulletin*, 21(10), 2953–2961.

Ahmad, U., Hussain, M., Ahmad, W., Javed, J., Arshad, Z., & Akram, Z. (2024). Impact of global climate change on maize (Zea mays): Physiological responses and modern breeding techniques. *Trends in Biotechnology and Plant Sciences*, 2(1), 62–77.

Ahmar, S., Gill, R. A., Jung, K. H., Faheem, A., Qasim, M. U., & Mubeen, M. et al. (2020). Conventional and molecular techniques from simple breeding to speed breeding in crop plants: Recent advances and future outlook. *International Journal of Molecular Sciences*, 21, 2590. doi: 10.3390/ijms21072590

Ahmed, C. B., Magdich, S., Rouina, B. B., Boukhris, M., & Abdullah, F. B. (2012). Saline water irrigation effects on soil salinity distribution and some physiological responses of field grown Chemlali olive. *Journal of Environmental Management*, 113, 538–544.

Ahmed, S., Iqbal, M., Ahmad, Z., Iqbal, M. A., Artyszak, A., & El-Sabagh, A. et al. (2023). Foliar application of silicon-based nanoparticles improve the adaptability of maize (*Zea mays* L.) in cadmium contaminated soils. *Environmental Science and Pollution Research*, 30, 41002–41013.

Ahmed, Z., Khalid, M., Ghafoor, A., Shah, M. K. N., Raja, G. K., Rana, R. M., Mahmood, T., & Thompson, A. M. (2022). SNP-based genome-wide association mapping of pollen viability under heat stress in tropical *Zea mays* L. inbred lines. *Frontiers in Genetics*, 13, 819849. doi: 10.3389/fgene.2022.819849

Ahuja, I., de Vos, R. C., Bones, A. M., & Hall, R. D. (2010). Plant molecular stress responses face climate change. *Trends in Plant Science*, 15, 664–674.

Ainley, W. M., Sastry-Dent, L., Welter, M. E., Murray, M. G., Zeitler, B., Amora, R., & Petolino, J. F. (2013). Trait stacking via targeted genome editing. *Plant Biotechnology Journal*, 11(9), 1126–1134.

Akbari, G., Sanavy, S. A., & Yousefzadeh, S. (2007). Effect of auxin and salt stress (NaCl) on seed germination of wheat cultivars (*Triticum aestivum* L.). *Pakistan Journal of Biological Sciences*, 10, 2557–2561.

Akram, M., Ashraf, M. Y., Ahmad, R., Rafiq, M., Ahmad, I., & Iqbal, J. (2010). Allometry and yield components of maize (Zea mays L.) hybrids to various potassium levels under saline conditions. *Archives of Biological Sciences*, 62(4), 1053–1061.

Aksoy, A., Celik, A., Ozturk, M., & Tulu, M. (2000). Roadside plants as possible indicators of heavy metal pollution in Turkey. *Chemia I Inzynieria Ekologiczna*, 7(11), 1152–1162.

Alam, M. A., Seetharam, K., Zaidi, P. H., Dinesh, A., Vinayan, M. T., & Nath, U. K. (2017). Dissecting heat stress tolerance in tropical maize (*Zea mays* L.). *Field Crops Research*, 204, 110–119. doi: 10.1016/j.fcr.2017.01.006

Alharbi, O. M., Khattab, R. A., & Ali, I. (2018). Health and environmental effects of persistent organic pollutants. *Journal of Molecular Liquids*, 263, 442–453.

Ali, B., Hayat, S., Fariduddin, Q., & Ahmad, A. (2008). 24-epibrassinolide protects against the stress generated by salinity and nickel in *Brassica juncea*. *Chemosphere*, 72, 1387–1392.

Ali, I., & Aboul-Enein, H. Y. (2002). Speciation of arsenic and chromium metal ions by reversed phase high performance liquid chromatography. *Chemosphere*, 48(3), 275–278.

Ali, I., Aboul-Enein, H. Y., & Gupta, V. K. (2009). *Nanochromatography and nanocapillary electrophoresis: Pharmaceutical and environmental analyses*. New York: John Wiley & Sons.

Ali, I., Alharbi, O. M., Alothman, Z. A., Badjah, A. Y., & Alwarthan, A. (2018). Artificial neural network modelling of amido black dye sorption on iron composite nano material: Kinetics and thermodynamics studies. *Journal of Molecular Liquids*, 250, 1–8.

Ali, I., Alothman, Z. A., & Al-Warthan, A. (2016b). Sorption, kinetics and thermodynamics studies of atrazine herbicide removal from water using iron nano-composite material. *International Journal of Environmental Science and Technology*, 13(2), 733–742.

Ali, I., & Jain, C. K. (2004). Advances in arsenic speciation techniques. *International Journal of Environmental Analytical Chemistry*, 84(12), 947–964.

Ali, I., Khan, T. A., & Asim, M. (2011). Removal of arsenic from water by electrocoagulation and electrodialysis techniques. *Separation & Purification Reviews*, 40(1), 25–42.

Ali, Q., Athar, H. U. R., & Ashraf, M. (2008). Modulation of growth, photosynthetic capacity and water relations in salt stressed wheat plants by exogenously applied 24-epibrassinolide. *Plant Growth Regulators*, 56, 107–116.

Aliu, S., Rusinovci, I., Fetahu, S., Gashi, B., Simeonovska, E., & Rozman, L. (2015). The effect of salt stress on the germination of maize (*Zea mays* L.) seeds and photosynthetic pigments. *Acta Agriculturae Slovenica*, 105, 85–94.

Allen, J. A. (1994). *Intraspecific variation in the response of baldcypress (Taxodium distichum) seedlings to salinity*, Ann Arbor, MI, Louisiana State University and Agricultural & Mechanical College.

Almeida, G. D., Makumbi, D., Magorokosho, C., Nair, S., Borém, A., & Ribaut, J. M. et al. (2013). QTL mapping in three tropical maize populations reveals a set of constitutive and adaptive genomic regions for drought tolerance. *Theoretical and Applied Genetics*, 126, 583–600. doi: 10.1007/s00122-012-2003-7

Amara, I., Capellades, M., Ludevid, M. D., Pagès, M., & Goday, A. (2013). Enhanced water stress tolerance of transgenic maize plants over-expressing LEA Rab28 gene. *Journal of Plant Physiology*, 170, 864–873. doi: 10.1016/j.jplph.2013.01.004

Anami, S., De Block, M., Machuka, J., & Van Lijsebettens, M. (2009). Molecular improvement of tropical maize for drought stress tolerance in sub-Saharan Africa. *Critical Reviews in Plant Sciences*, 28, 16–35. doi: 10.1080/07352680802665305

Andersson, M., Turesson, H., Olsson, N., Fält, A. S., Ohlsson, P., & Gonzalez, M. N. et al. (2018). Genome editing in potato via CRISPR-Cas9 ribonucleoprotein delivery. *Physiologia Plantarum*, 164, 378–384. doi: 10.1111/ppl.12731

Andjelković, V., & Ignjatović-Micić, D. (2011). ESTs analysis in maize developing kernels exposed to single and combined water and heat stresses. *Scientia Agricola*, 68, 353–360. doi: 10.1590/s0103-90162011000300013

Armstrong, W., & Drew, M. C. (2002). Root growth and metabolism under oxygen deficiency. In: A. Eshel & U. Kafkafi, eds. *Plant roots: The hidden half* (3rd ed., pp. 729–761). New York, USA: Marcel Dekker,.

Arnold, C. Y. (1974). Predicting stages of sweet corn (*Zea mays* L.) development. *Journal American Society for Horticultural Science*, 99(6), 501–505. doi: 10.21273/JASHS.99.6.501

Arora, N., Bhardwaj, R., Sharma, P., & Arora, H. K. (2008). 28-Homobrassinolide alleviates oxidative stress in salt treated maize (*Zea mays* L.) plants. *Brazilian Journal of Plant Physiology*, 20, 153–157.

Ashraf, M. (2010). Inducing drought tolerance in plants: Recent advances. *Biotechnology Advances*, 28, 169–183. doi: 10.1016/j.biotechadv.2009.11.005

Ashraf, M., Ozturk, M., & Ahmad, M. S. A., eds. (2010a). *Plant adaptation and phytoremediation*, New York: Springer, 481.

Ashraf, M., & Rauf, H. (2001). Inducing salt tolerance in maize (*Zea mays* L.) through seed priming with chloride salts: Growth and ion transport at early growth stages. *Acta Physiologiae Plantarum*, 23, 407–414.

Ashrafuzzaman, M., Khan, M. H., & Shahidullah, S. M. (2002). Vegetative growth of maize (*Zea mays*) as affected by a range of salinity. *Crop Research*, 24, 286–291.

Aslam, M., Maqbool, M. A., & Cengiz, R. (2015). *Drought stress in maize (Zea mays L.) effects, resistance mechanisms, global achievements*. Cham: Springer.

Aydinalp, C., & Marinova, S. (2009). The effects of heavy metals on seed germination and plant growth on alfalfa plant (*Medicago sativa*). *Bulgarian Journal of Agricultural Science*, 15(4), 347–350.

Aydinoglu, F. (2020). Elucidating the regulatory roles of microRNAs in maize (Zea mays L.) leaf growth response to chilling stress. *Planta*, 251, 38. doi: 10.1007/s00425-019-03331-y

Azevedo Neto, A. D., Prisco, J. T., Enéas-Filho, J., Lacerda, C. F., Silva, J. V., & Costa, P. H. et al. (2004). Effects of salt stress on plant growth, stomatal response and solute accumulation of different maize genotypes. *Brazilian Journal of Plant Physiology*, 16, 31–38.

Bahuguna, R. N., & Jagadish, K. S. V. (2015). Temperature regulation of plant phenological development. *Environmental and Experimental Botany*, 111, 83–90. doi: 10.1016/j.envexpbot.2014.10.007

Bakht, J., Shafi, M., Jamal, Y., & Sher, H. (2011). Response of maize (*Zea mays* L.) to seed priming with NaCl and salinity stress. *Spanish Journal of Agricultural Research*, 9, 252–261.

Bakhtavar, M. A., Afzal, I., Basra, S. M. A., Ahmad, A. U. H., & Noor, M. A. (2015). Physiological strategies to improve the performance of spring maize (Zea mays L.) planted under early and optimum sowing conditions. *PloS One*, 10(4), e0124441.

Banerjee, S., Sirohi, A., Ansari, A. A., & Gill, S. S. (2017). Role of small RNAs in abiotic stress responses in plants. *Plant Gene*, 11, 180–189. doi: 10.1016/j.plgene.2017.04.005

Bravo-F, P., & Uribe, E. G. (1981). Temperature dependence of the concentration kinetics of absorption of phosphate and potassium in corn roots. *Plant Physiology*, 67(4), 815–819.

Barconi, D., Bernardini, G., & Santucci, A. (2011). Linking protein oxidation to environmental pollutants: Redox proteome approaches. *Journal of Proteomics*, 74(11), 2324–2337.

Barošević, T., Bagi, F., Savić, Z., Ljubičić, N., & Ivanović, I. (2022). Assessment of maize hybrids resistance to *Aspergillus* ear rot and aflatoxin production in environmental conditions in Serbia. *Toxins*, 14(12), 887. doi: 10.3390/toxins14120887

Basheer, A. A. (2018a). Chemical chiral pollution: Impact on the society and science and need of the regulations in the 21st century. *Chirality*, 30(4), 402–406.

Bastías, E. I., González-Moro, M. B., & González-Murua, C. (2004). Zea mays L. amylacea from the Lluta valley (Arica-Chile) tolerates salinity stress when high levels of boron are available. *Plant Soil*, 267, 73–84.

Battisti, D. S., & Naylor, R. L. (2009). Historical warnings of future food insecurity with unprecedented seasonal heat. *Science*, 323, 240–244.

Beló, A., Zheng, P., Luck, S., Shen, B., Meyer, D. J., & Li, B. et al. (2008). Whole genome scan detects an allelic variant of fad2 associated with increased oleic acid levels in maize. *Molecular Genetics & Genomics*, 279, 1–10. doi: 10.1007/s00438-007-0289-y

Benešová, M., Hola, D., Fischer, L., Jedelský, P. L., Hnilièka, F., & Wilhelmová, N. et al. (2012). The physiology and proteomics of drought tolerance in maize: Early stomatal closure as a cause of lower tolerance to short-term dehydration? *PLoS One*, 7, e38017. doi: 10.1371/journal.pone.0038017

Berberich, T., Sano, H., & Kusano, T. (1999). Involvement of a MAP kinase, ZmMPK5, in senescence and recovery from low-temperature stress in maize. *Molecular & General Genetics*, 262, 534–542. doi: 10.1007/s004380051115.

Bernardo, R. (2008). Molecular markers and selection for complex traits in plants: Learning from the last 20 years. *Crop Science*, 48, 1649–1664.

Beyene, Y., Semagn, F. K., Mugo, S. N., Tarekegne, A. T., Babu, R., & Meisel, B. et al (2015). Genetic gains in grain yield through genomic selection in eight bi-parental maize populations under drought stress. *Crop Science*, 55, 154–163. doi: 10.2135/cropsci2014.07.0460

Beyene, Y., Semagn, K., Crossa, J., Mugo, S., Atlin, G., & Tarekegne, A. T. et al. (2016). Improving maize grain yield under drought stress and non-stress environments in sub-Saharan Africa using marker-assisted recurrent selection. *Crop Science*, 56, 344–353. doi: 10.2135/cropsci2015.02.0135

Bhat, M. A., Mir, R. A., Kumar, V., Shah, A. A., Zargar, S. M., Rahman, S., & Jan, A. T. (2021). Mechanistic insights of CRISPR/Cas mediated genome editing towards enhancing abiotic stress tolerance in plants. *Physiologia Plantarum*, 172(2), 1255–1268. doi: 10.1111/ppl.13359

Bakhtavar, M. A., Afzal, I., Basra, S. M. A., Ahmad, A. U. H., & Noor, M. A. (2015). Physiological strategies to improve the performance of spring maize (Zea mays L.) planted under early and optimum sowing conditions. *PloS One*, 10(4), e0124441.

Bojović, B., Đelić, G., Topuzović, M., & Stanković, M. (2010). Effects of NaCl on seed germination in some species from families Brassicaceae and Solanaceae. *Kragujevac Journal of Science*, 32, 83–87.

Bonhomme, L., Valot, B., Tardieu, F., & Zivy, M. (2012). Phosphoproteome dynamics upon changes in plant water status reveal early events associated with rapid growth adjustment in maize leaves. *Molecular & Cellular Proteomics*, 11, 957–972. doi: 10.1074/mcp.M111.015867

Borlaug, N. (2007). Feeding a hungry world. *Science*, 318(5849), 359.

Bu, L. D., Liu, J. L., Zhu, L., Luo, S. S., Chen, X. P., Li, S. Q., & Zhao, Y. (2013). The effects of mulching on maize growth, yield and water use in a semi-arid region. *Agricultural Water Management*, 123, 71–78.

Budak, H., Hussain, B., Khan, Z., Ozturk, N. Z., & Ullah, N. (2015). From genetics to functional genomics: improvement in drought signaling and tolerance in wheat. *Frontiers in Plant Science*, 6, 1012. doi: 10.3389/fpls.2015.01012

Burakova, E. A., Dyachkova, T. P., Rukhov, A. V., Tugolukov, E. N., Galunin, E. V., Tkachev, A. G., & Ali, I. (2018). Novel and economic method of carbon nanotubes synthesis on a nickel magnesium oxide catalyst using microwave radiation. *Journal of Molecular Liquids*, 253, 340–346.

Burke, M. B., Lobell, D. B., & Guarino, L. (2009). Shifts in African crop climates by 2050, and the implications for crop improvements and genetic resources conservation. *Global Environmental Change*, 19(3), 317–325.

Bybordi, A. (2010). The influence of salt stress on seed germination, growth and yield of canola cultivars. *Notulae Botanicae Horti Agrobotanici Cluj-Napoca*, 38, 128–133.

Cabot, C., Sibole, J. V., Barceló, J., & Poschenrieder, C. (2009). Abscisic acid decreases leaf Na+ exclusion in salt-treated Phaseolus vulgaris L. *Journal of Plant Growth Regulation*, 28(2), 187–192.

Cai, G., Wang, G., Wang, L., Pan, J., Liu, Y., & Li, D. (2014). ZmMKK1, a novel group A mitogen-activated protein kinase kinase gene in maize, conferred chilling stress tolerance and was involved in pathogen defense in transgenic tobacco. *Plant Science*, 214, 57–73. doi: 10.1016/j.plantsci.2013.09.014

Cairns, J. E., Hellin, J., Sonder, K., Araus, J. L., MacRobert, J. F., Thierfelder, C., & Prasanna, B. (2013). Adapting maize production to climate change in sub-Saharan Africa. *Food Security*, 5, 345–360.

Cantarero, M. G., Cirilo, A. G., & Andrade, F. H. (1999). Night temperature at silking affects set in maize. *Crop Science*, 39(3), 703–710. https://doi.org/10.2135/cropsci1999.0011183X003900020017xDigital Object Identifier (DOI)

Cao, S., Xu, Q., Cao, Y., Qian, K., An, K., & Zhu, Y. et al. (2005). Loss of-function mutations in DET2 gene lead to an enhanced resistance to oxidative stress in Arabidopsis. *Physiologia Plantarum*, 123, 57–66.

Cao, Y., Zhou, X., Song, H., Zhang, M., & Jiang, C. (2023). Advances in deciphering salt tolerance mechanism in maize. *The Crop Journal*, 11, 1001–1010.

Capelle, V., Remoue, C., Moreau, L., Reyss, A., Mahé, A., & Massonneau, A. et al. (2010). QTLs and candidate genes for desiccation and abscisic acid content in maize kernels. *BMC Plant Biology*, 10(2). doi: 10.1186/1471-2229-10-2

Carberry, P. S., Muchow, R. C., & Mccown, R. L. (1989). Testing the CERES-Maize simulation model in a semi-arid tropical environment. *Field Crops Research*, 20(4), 297–315. doi: 10.1016/0378-4290(89)90072-5

Carneiro, A. K., Montessoro, P. D. F., Fusaro, A. F., Araújo, B. G., & Hemerly, A. S. (2021). Plant CDKs—driving the cell cycle through climate change. *Plants*, 10, 1804. doi: 10.3390/plants10091804

Carpici, E. B., Celik, N., & Bayram, G. (2010). The effects of salt stress on the growth, biochemical parameter and mineral element content of some maize (Zea mays L.) cultivars. *African Journal of Biotechnology*, 9(41), 6937–6942.

Carrara, M., Prischi, F., Nowak, P. R., & Ali, M. M. (2015). Crystal structures reveal transient PERK luminal domain tetramerization in endoplasmic reticulum stress signaling. *The EMBO Journal*, 34(11), 1589–1600. doi: 10.15252/embj.201489183

Carter, E. K., Melkonian, J., Riha, S. J., & Shaw, S. B. (2016). Separating heat stress from moisture stress: Analyzing yield response to high temperature in irrigated maize. *Environmental Research Letters*, 11(9), 094012. doi: 10.1088/1748-9326/11/9/094012

Carvalho, R. F., Piotto, F. A., Schmidt, D., Peters, L. P., Monteiro, C. C., & Azevedo, R. A. (2011). Seed priming with hormones does not alleviate induced oxidative stress in maize seedlings subjected to salt stress. *Scientia Agricola*, 68, 598–602.

Casaretto, J. A., El-Kereamy, A., Zeng, B., Stiegelmeyer, S. M., Chen, X., Bi, Y.-M., & Rothstein, S. J. (2016). Expression of OsMYB55 in maize activates stress-responsive genes and enhances heat and drought tolerance. *BMC Genomics*, 17(1), 312. doi: 10.1186/s12864-016-2659-5

Celik, S., Yucel, E., Celik, S., Gucel, S., & Ozturk, M. (2010). Carolina poplar (Populus x canadensis Moench) as a biomonitor of trace elements in Black Sea region of Turkey. *Journal of Environmental Biology*, 31(1), 225.

Cermák, T., Baltes, N. J., Čegan, R., Zhang, Y., & Voytas, D. F. (2015). High-frequency, precise modification of the tomato genome. *Genome Biology*, 16, 232. doi: 10.1186/s13059-015-0796-9

Chakrabarti, N., & Mukherji, S. (2003). Alleviation of NaCl stress by pretreatment with phytohormones in Vigna radiata. *Biologia Plantarum*, 46(4), 589–594.

Chakraborty, D., Sehgal, V. K., Dhakar, R., Ray, M., & Das, D. K. (2019). Spatio-temporal trend in heat waves over India and its impact assessment on wheat crop. *Theoretical and Applied Climatology*, 138, 1925–1937. doi: 10.1007/s00704-019-02939-0

Char, S. N., Unger-Wallace, E., Frame, B., Briggs, S. A., Main, M., Spalding, M. H., Vollbrecht, E., Wang, K., & Yang, B. (2015). Heritable site-specific mutagenesis using TALENs in maize. *Plant Biotechnology Journal*, 13(7), 1002–1010. doi: 10.1111/pbi.12344

Chattha, M. U., Khan, M. A., Khan, I., Mahmood, A., Chattha, M. B., & Hassan, M. U. et al. (2023). Comparison of physio-biochemical and antioxidant enzymes in maize during early growth stage in response to salt stress. *Pakistan Journal of Botany*, 55, 1991–1997.

Chaumont, F., Barrieu, F., Wojcik, E., Chrispeels, M. J., & Jung, R. (2001). Aquaporins constitute a large and highly divergent protein family in maize. *Plant Physiology*, 125, 1206–1215.

Chen, L. J., Li, C. S., Feng, Q., Wei, Y. P., Zheng, H., & Zhao, Y. et al. (2017). Shifts in soil microbial metabolic activities and community structures along a salinity gradient of irrigation water in a typical arid region of China. *Science of the Total Environment*, 598, 64–70.

Chen, M., & Tang, Y. (2012). Effects of low temperature stress on chlorophyll fluorescence characteristics of com seedlings. *Journal of Inner Mongolia Agricultural University (Natural Science Edition)*, 33, 20–24. doi: 10.16853/j.cnki.1009-3575.2012.03.006

Chen, X., Gu, Z., Xin, D., Hao, L., Liu, C., Huang, J., Ma, B., & Zhang, H. (2011). Identification and characterization of putative CIPK genes in maize. *Journal of Genetics and Genomics*, 38(2), 77–87. doi: 10.1016/j.jcg.2011.01.005

Chen, Y., Cao, Y., Wang, L., Li, L., Yang, J., & Zou, M. (2018). Identification of MYB transcription factor genes and their expression during abiotic stresses in maize. *Biologia Plantarum*, 62, 222–230. doi: 10.1007/s10535-017-0756-1

Choudhary, S. K., Kumar, V., Singhal, R. K., Bose, B., Chauhan, J., & Alamri, S. et al. (2021). Seed priming with Mg(NO3)2 and ZnSO4 salts triggers the germination and growth attributes synergistically in wheat varieties. *Agronomy*, 11, 2110.

Chukwudi, U. P., Kutu, F. R., & Mavengahama, S. (2021). Heat stress effect on the grain yield of three drought-tolerant maize varieties under varying growth conditions. *Plants (Basel)*, *10*(8), 1532. PMID: 34451577; PMCID: PMC8401389. doi: 10.3390/plants10081532

Cohen, S. P., & Leach, J. E. (2020). High temperature-induced plant disease susceptibility: More than the sum of its parts. *Current Opinion in Plant Biology*, 56, 235–241. doi: 10.1016/j.pbi.2020.02.008

Commuri, P. D., & Jones, R. J. (2001). High temperatures during endosperm cell division in maize: A genotypic comparison under in vitro and field conditions. *Crop Science*, 41(4), 1122–1130. doi: 10.2135/cropsci2001.4141122x

Cortijo, S., Wardenaar, R., Colomé-Tatché, M., Gilly, A., Etcheverry, M., & Labadie, K. et al. (2014). Mapping the epigenetic basis of complex traits. *Science*, 343, 1145. doi: 10.1126/science.1248127

Coşkun, Y., Coskun, A., Demirel, U., & Ozden, M. (2011). Physiological response of maize (*Zea mays* L.) to high temperature stress. *Australian Journal of Crop Science*, 5(8), 966–972.

Crafts-Brandner, S. J., & Salvucci, M. E. (2002). Sensitivity of photosynthesis in a C4 plant, maize, to heat stress. *Plant Physiology*, 129(4), 1773–1780. doi: 10.1104/pp.002170

Daghan, H., Öztürk, M., Hakeem, K., Sabir, M., & Mermut, A. (2015). Soil pollution in Turkey and remediation methods. In: K. R. Hakeem, M. Sabir, M. Ozturk, & A. R. Mermut, eds. *Soil remediation and plants prospects and challenges* (pp. 287–312). London: Academic Press.

Daničić, M., Pejić, B., Mačkić, K., Lalić, B., Maksimović, I., & Putnik-Delić, M. (2021). The predicted impact of climate change on maize production in Northern Serbia. *Maydica*, 65(3), 10.

Darra, B. L., & Saxena, S. N. (1973). Role of IAA on the mineral composition of maize (Zea mays) crop under various osmotic stressed conditions. *Plant and Soil*, 38(3), 657–661.

Davani, D., Nabipour, M., & Roshanfekr Dezfouli, H. (2017). Effects of different concentrations of cytokinin and auxin hormones on yield and yield components of grain maize (Zea mays L.) in salinity conditions. *Plant Productions*, 40(1), 69–80.

de Azevedo Neto, A. D., Prisco, J. T., Enéas-Filho, J., Medeiros, J. V. R., & Gomes-Filho, E. (2005). Hydrogen peroxide pre-treatment induces salt-stress acclimation in maize plants. *Journal of Plant Physiology*, 162(10), 1114–1122.

de Freitas, P. A. F., de Souza Miranda, R., Marques, E. C., Prisco, J. T., & Gomes Filho, E. (2018). Salt tolerance induced by exogenous proline in maize is related to low oxidative damage and favorable ionic homeostasis. *Journal of Plant Growth Regulation*, 37, 911–924.

Dehghani, M. H., Sanaei, D., Ali, I., & Bhatnagar, A. (2016). Removal of chromium (VI) from aqueous solution using treated waste newspaper as a low-cost adsorbent: Kinetic modeling and isotherm studies. *Journal of Molecular Liquids*, 215, 671–679.

Deikman, J., Petracek, M., & Heard, J. E. (2012). Drought tolerance through biotechnology: Improving translation from the laboratory to farmers' fields. *Current Opinion in Biotechnology*, 23, 243–250. doi: 10.1016/j.copbio.2011.11.003

Deng, C. N., Zhang, G. X., Pan, X. L., & Zhao, K. Y. (2010). Chlorophyll fluorescence and gas exchange responses of maize seedlings to saline-alkaline stress. *Bulgarian Journal of Agricultural Science*, 16, 49–58.

Dhankher, O. P., & Foyer, C. H. (2018). Climate resilient crops for improving global food security and safety. *Plant, Cell & Environment*, 41(5), 877–884. doi: 10.1111/pce.13207

Di Fenza, M., Hogg, B., Grant, J., & Barth, S. (2017). Transcriptomic response of maize primary roots to low temperatures at seedling emergence. *PeerJ*, 5, e2839. doi: 10.7717/peerj.2839

Dinka, S. J., Campbell, M. A., Demers, T., & Raizada, M. N. (2007). Predicting the size of the progeny mapping population required to positionally clone a gene. *Genetics*, 176, 2035–2054. doi: 10.1534/genetics.107.074377

Djalovic, I., Kundu, S., Bahuguna, R. N., Pareek, A., Raza, A., Singla-Pareek, S. L., Prasad, P. V. V., & Varshney, R. K. (2024). Maize and heat stress: Physiological, genetic, and molecular insights. *The Plant Genome*, 17, e20378. doi: 10.1002/tpg2.20378

Djanaguiraman, M., Senthil, A., & Ramadass, R. (2003). Assessment of rice genotypes for salinity tolerance at germination and seedling stage. *Madras Agricultural Journal*, 90, 506–510.

Dong, A., Yang, Y., Liu, S., Zenda, T., Liu, X., & Wang, Y. et al. (2020). Comparative proteomics analysis of two maize hybrids revealed drought-stress tolerance mechanisms. *Biotechnology & Biotechnological Equipment*, 34, 763–780. doi: 10.1080/13102818.2020.1805015

Dong, X., Guan, L., Zhang, P., Liu, X., Li, S., Fu, Z., Tang, L., Qi, Z., Qiu, Z., Jin, C., Huang, S., & Yang, H. (2021). Responses of maize with different growth periods to heat stress around flowering and early grain filling. *Agricultural and Forest Meteorology*, 303, 108378.

Dubey, A., Kumar, A., & Khan, M. L. (2020). Role of biostimulants for enhancing abiotic stress tolerance in Fabaceae plants. In: *The plant family fabaceae: Biology and physiological responses to environmental stresses* (pp. 223–236). Singapore: Springer Singapore.

Duran Garzon, C., Lequart, M., Rautengarten, C., Bassard, S., Sellier-Richard, H., & Baldet, P. et al. (2020). Regulation of carbon metabolism in two maize sister lines contrasted for chilling tolerance. *Journal of Experimental Botany*, 71, 356–369. doi: 10.1093/jxb/erz421

Edmeades, G. O., Banziger, M., & Beck, D. (1997). Development and per se performance of CIMMYT maize populations as drought tolerant sources. In: G. O. Edmeades, ed.

Edmeades, G. O., Bolaños, J., Hernandez, M., & Bello, S. (1993). Causes for silk delay in a lowland tropical maize population. *Crop Science*, 33(5), 1029–1035.

El Haddad, N., Choukri, H., Ghanem, M. E., Smouni, A., Mentag, R., Rajendran, K., Hejjaoui, K., Maalouf, F., & Kumar, S. High-Temperature and drought stress effects on growth, yield and nutritional quality with transpiration response to vapor pressure deficit in lentil. *Plants*, 11, 95. doi: 10.3390/plants11010095

Elhakem, A. H. (2020). Salicylic acid ameliorates salinity tolerance in maize by regulation of phytohormones and osmolytes. *Plant, Soil and Environment*, 66, 533–541.

El-Sappah, A. H., Rather, S. A., Wani, S. H., Elrys, A. S., Bilal, M., Huang, Q., Dar, Z. A., Elashtokhy, M. M. A., Soaud, N., Koul, M., Mir, R. R., Yan, K., Li, J., El-Tarabily, K. A., & Abbas, M. (2022). Heat stress-mediated constraints in maize (*Zea mays*) production: Challenges and solutions. *Frontiers in Plant Science*, 13, 879366. doi: 10.3389/fpls.2022.879366

El Sayed, H. E., & El Sayed, A. (2011). Influence of salinity stress on growth parameters, photosynthetic activity and cytological studies of *Zea mays*, L. plant using hydrogel polymer. *Agriculture and Biology Journal of North America*, 2, 907–920.

El-Sharkawy, H. M., Shehata, S. A., Eisa, S., Kishk, E. T., Khafaga, H. S., & El-Naby, A. S. A. (2017). Foliar application of thidiazuron, potassium chloride and boron with early cultivation date elevated growth and productivity of hybrid corn grown under adverse conditions. *International Journal of Environment*, 6, 31–41.

Emamverdian, A., Ding, Y., Mokhberdoran, F., & Xie, Y. (2015). Heavy metal stress and some mechanisms of plant defense response. *The Scientific World Journal*, 2015, 1–18.

Eremina, M., Rozhon, W., & Poppenberger, B. (2016). Hormonal control of cold stress responses in plants. *Cellular and Molecular Life Sciences*, 73, 797–810. doi: 10.1007/s00018-015-2089-6

Erenstein, O., Jaleta, M., Sonder, K., Mottaleb, K., & Prasanna, B. M. (2022). Global maize production, consumption and trade: Trends and R&D implications. *Food Security*, 14(5), 1295–1319.

Erpen, L., Devi, H. S., Grosser, J. W., & Dutt, M. (2018). Potential use of the DREB/ERF, MYB, NAC and WRKY transcription factors to improve abiotic and biotic stress in transgenic plants. *Plant Cell, Tissue and Organ Culture*, 132, 1–25. doi: 10.1007/s11240-017-1320-6

Evlagon, D., Ravina, I., & Neumann, P. M. (1992). Effects of salinity stress and calcium on hydraulic conductivity and growth in maize seedling roots. *Journal of Plant Nutrition*, 15, 795–803.

Eyidogan, F., Oz, M. T., Yucel, M., & Oktem, H. A. (2012). Signal transduction of phytohormones under abiotic stresses. In: N. Khan, R. Nazar, N. Iqbal, & N. Anjum, eds. *Phytohormones and abiotic stress tolerance in plants*. Springer. doi: 10.1007/978-3-642-25829-9_1

FAO (2021). *The impact of disasters and crises on agriculture and food security*. Rome: Food and agriculture organization of the United Nations.

FAOStat. (2021). *FAO stat*. FAO. http://www.fao.org/faostat

Faostat FJQ (2017) Available online: http://www.fao.org/faostat/en/#data. Accessed Dec 2019.

Farfan, I. D. B., De La Fuente, G. N., Murray, S. C., Isakeit, T., Huang, P. C., Warburton, M., ... & Kolomiets, M. (2015). Genome wide association study for drought, aflatoxin resistance, and important agronomic traits of maize hybrids in the sub-tropics. *PloS One*, 10(2), e0117737.

Farhangi-Abriz, S., & Torabian, S. (2017). Antioxidant enzyme and osmotic adjustment changes in bean seedlings as affected by biochar under salt stress. *Ecotoxicology and Environmental Safety*, 137, 64–70. doi: 10.1016/j.ecoenv.2016.11.029

Farooq, M., Aziz, T., Wahid, A., Lee, D.-J., & Siddique, K. H. (2009). Chilling tolerance in maize: Agronomic and physiological approaches. *Crop & Pasture Science*, 60, 501–516. doi: 10.1071/CP08427

Farooq, M., Basra, S. M., Rehman, H., & Mehmood, T. (2006). Germination and early seedling growth as affected by pre-sowing ethanol seed treatments in fine rice. *International Journal of Agriculture and Biology*, 8, 19–22.

Farooq, M. S., Uzair, M., Raza, A., Habib, M., Xu, Y., Yousuf, M., Yang, S. H., & Khan, M. R. (2022). Uncovering the research gaps to alleviate the negative impacts of climate change on food security: A review. *Frontiers in Plant Science*, 13, 927535. doi: 10.3389/fpls.2022.927535

Farsiani, A., & Ghobadi, M. E. (2009). Effects of PEG and NaCl stress on two cultivars of corn (Zea mays L.) at germination and early seedling stages. *World Academy of Science, Engineering and Technology*, 57, 382–385.

Fauser, F., Schiml, S., & Puchta, H. (2014). Both CRISPR/Cas-based nucleases and nickases can be used efficiently for genome engineering in Arabidopsis thaliana. *The Plant Journal*, 79, 348–359. doi: 10.1111/tpj.12554

Fedoroff, N. V. et al. (2010). Radically rethinking agriculture for the 21st century. *Science*, 327, 833–834.

Finka, A., Cuendet, A. F. H., Maathuis, F. J. M., Saidi, Y., & Goloubinoff, P. (2012). Plasma membrane cyclic nucleotide gated calcium channels control land plant thermal sensing and acquired thermotolerance. *Plant Cell*, 24, 3333–3348. doi: 10.1105/tpc.112.095844

Fita, A., Rodríguez-Burruezo, A., Boscaiu, M., Prohens, J., & Vicente, O. (2015). Breeding and domesticating crops adapted to drought and salinity: A new paradigm for increasing food production. *Frontiers in Plant Science*, 6, 978. doi: 10.3389/fpls.2015.00978

Fracheboud, Y., Ribaut, J. M., Vargas, M., Messmer, R., & Stamp, P. (2002). Identification of quantitative trait loci for cold-tolerance of photosynthesis in maize (Zea mays L.). *Journal of Experimental Botany*, 53, 1967–1977. doi: 10.1093/jxb/erf040

Frascaroli, E., & Revilla, P. (2018). Genomics of cold tolerance in maize. In: J. Bennetzen, S. Flint-Garcia, C. Hirsch, & R. Tuberosa, eds. *The maize genome*. Cham: Springer International Publishing, 287–303. doi: 10.1007/978-3-319-97427-9_17

Freitas, V. S., de Souza Miranda, R., Costa, J. H., de Oliveira, D. F., de Oliveira Paula, S., de Castro Miguel, E., ... & Gomes-Filho, E. (2018). Ethylene triggers salt tolerance in maize genotypes by modulating polyamine catabolism enzymes associated with H2O2 production. *Environmental and Experimental Botany*, 145, 75–86.

Frey, F. P., Pitz, M., Schön, C. C., & Hochholdinger, F. (2020). Transcriptomic diversity in seedling roots of European flint maize in response to cold. *BMC Genomics*, 21(1), 300. doi: 10.1186/s12864-020-6682-1

Frova, C., Krajewski, P., Di Fonzo, N., Villa, M., & Sari-Gorla, M. (1999). Genetic analysis of drought tolerance in maize by molecular markers II. Plant height and flowering. *Theoretical and Applied Genetics*, 99, 289–295. doi: 10.1007/s001220051234

Fu, J., Cheng, Y., Linghu, J., Yang, X., Kang, L., & Zhang, Z. et al. (2013). RNA sequencing reveals the complex regulatory network in the maize kernel. *Nature Communications*, 4, 2832. doi: 10.1038/ncomms3832

Fu, J., Momčilović, I., & Prasad, P. V. V. (2012). Roles of protein synthesis elongation factor EF-Tu in heat tolerance in plants. *Journal of Botany*, 2012, 835836. doi: 10.1155/2012/835836

Gabaldón-Leal, C., Webber, H., Otegui, M. E., Slafer, G. A., Ordóñez, R. A., Gaiser, T., Lorite, I. J., Ruiz-Ramos, M., & Ewert, F. (2016). Modelling the impact of heat stress on maize yield formation. *Field Crops Research*, 198, 226–237. doi: 10.1016/j.fcr.2016.08.013

Gadalla, A. M., Hamdy, A., Galal, Y. G. M., Aziz, H. A., & Mohamed, M. A. A. (2007). Evaluation of maize grown under salinity stress and N application strategies using stable nitrogen isotope.

Gaffney, I., Sallach, J. B., Wilson, J., Bergström, E., & Thomas-Oates, J. (2021). Metabolomic approaches to studying the response to drought stress in corn (Zea mays) cobs. *Metabolites*, 11, 438. doi: 10.3390/metabo11070438

Gao, F., Han, X., Wu, J., Zheng, S., Shang, Z., Sun, D., Zhou, R., & Li, B. (2012). A heat-activated calcium-permeable channel—Arabidopsis cyclic nucleotide-gated ion channel 6 - Is involved in heat shock responses. *Plant Journal*, 70(6), 1056–1069. doi: 10.1111/j.1365-313X.2012.04969.x

Gao, J., Yang, Y., Wang, H., Wang, P., Li, H., Li, M., Ren, L., Yue, X., & Liao, H. (2021). Fast climate responses to emission reductions in aerosol and ozone precursors in China during 2013–2017. *Atmospheric Chemistry and Physics*, 22, 7131–7142.

Gawad, G. A., Arslan, A., Gaihbe, A., & Kadouri, F. (2005). The effects of saline irrigation water management and salt tolerant tomato varieties on sustainable production of tomato in Syria (1999-2002). *Agricultural Water Management*, 78, 39–53.

Genchi, G., Sinicropi, M. S., Lauria, G., Carocci, A., & Catalano, A. (2020). The effects of cadmium toxicity. *International Journal of Environmental Research and Public Healt h*, 17(11), 3782. doi: 10.3390/ijerph17113782

GISTEMP Team. (2023). *GISS Surface Temperature Analysis (GISTEMP)*, version 4. NASA Goddard Institute for Space Studies. https://data.giss.nasa.gov/gistemp/

Gómez-Muñoz, B., Lekfeldt, J. D., Magid, J., Jensen, L. S., & de Neergaard, A. (2018). Seed treatment with penicillium sp. or Mn/Zn can alleviate the negative effects of cold stress in maize grown in soils dependent on soil fertility. *Journal of Agronomy and Crop Science*, 204, 603–612. doi: 10.1111/jac.12288

Gondim, F. A., Gomes-Filho, E., Costa, J. H., Alencar, N. L. M., & Prisco, J. T. (2012). Catalase plays a key role in salt stress acclimation induced by hydrogen peroxide pretreatment in maize. *Plant Physiology and Biochemistry*, 56, 62–71.

Gondim, F. A., Gomes-Filho, E., Lacerda, C. F., Prisco, J. T., Neto, A. D. A., & Marques, E. C. (2010). Pretreatment with H2O2 in maize seeds: Effects on germination and seedling acclimation to salt stress. *Brazilian Journal of Plant Physiology*, 22, 103–112.

Gong, D. H., Wang, G. Z., Si, W. T., Zhou, Y., Liu, Z., & Jia, J. (2018). Effects of salt stress on photosynthetic pigments and activity of ribulose-1, 5-bisphosphate carboxylase/oxygenase in Kalidium Foliatum. *Russian Journal of Plant Physiology*, 65, 98–103.

Gong, F., Yang, L., Tai, F., Hu, X., & Wang, W. (2014). "Omics" of maize stress response for sustainable food production: Opportunities and challenges. *Omics: A Journal of Integrative Biology*, 18, 714–732. doi: 10.1089/omi.2014.0125

Granier, C., Inze, D., & Tardieu, F. (2000). Spatial distribution of cell division rate can be deduced from that of p34cdc2 kinase activity in maize leaves grown at contrasting temperatures and soil water conditions. *Plant Physiology*, 124, 1393–1402. doi: 10.1104/pp.124.3.1393

Grewal, H. S., Norrish, S., & Cornish, P. (2004). Subsoil salts affect root function, shoot growth and ionic balance of wheat plants. New directions for a diverse planet. Proceedings of the 4th International Crop Science Congress; 2004 September 26-October 1, Brisbane, Australia. Berlin, Germany: ResearchGate.

Guan, Y., Hu, J., Wang, X., & Shao, C. (2009). Seed priming with chitosan improves maize germination and seedling growth in relation to physiological changes under low temperature stress. *Journal of Zhejiang University Science B*, 10, 427–433. doi: 10.1631/jzus.B0820373

Gücel, S., Kocbas, F., & Ozturk, M. (2009b). Metal bioaccumulation by barley in Mesaoria Plain alongside the Nicosia Famagusta Highway, Northern Cyprus. *Fresenius Environmental Bulletin*, 18(11), 2034–2039.

Gunes, A., Inal, A., Alpaslam, M., Erslan, F., Bagsi, E. G., & Cicek, N. (2007). Salicylic acid induced changes on some physiological parameters symptomatic for oxidative stress and mineral nutrition in maize (*Zea mays* L.) grown under salinity. *Journal of Plant Physiology*, 164, 728–736.

Gupta, V. K., & Ali, I. (2002). *Encyclopedia of surface and colloid science*, New York, Marcel Dekker, 136–166.

Gupta, V. K., & Ali, I. (2012). *Environmental water: advances in treatment, remediation and recycling*. London: Newnes.

Ha, S., Vankova, R., Yamaguchi-Shinozaki, K., Shinozaki, K., & Tran, L. S. (2012). Cytokinins: Metabolism and function in plant adaptation to environmental stresses. *Trends in Plant Science*, 17, 172–179.

Habben, J. E., Bao, X., Bate, N. J., DeBruin, J. L., Dolan, D., & Hasegawa, D. et al (2014). Transgenic alteration of ethylene biosynthesis increases grain yield in maize under field drought-stress conditions. *Plant Biotechnology Journal*, 12, 685–693. doi: 10.1111/pbi.12172

Hager, A. (2003). Role of the plasma membrane H^+-ATPase in auxin-induced elongation growth: Historical and new aspects. *Journal of Plant Research*, 116, 483–505.

Haider, S., Raza, A., Iqbal, J., Shaukat, M., & Mahmood, T. (2022). Analyzing the regulatory role of heat shock transcription factors in plant heat stress tolerance: A brief appraisal. *Molecular Biology Reports*, 49, 5771–6785. doi: 10.1007/s11033-022-07190-x

Hall, A. E. (2011). Breeding cowpea for future climates. In: S. S. Yadav, R. J. Redden, J. L. Hatfield, H. Lotez-Campen, & A. E. Hall, eds., *Crop adaptation to climate change* (pp. 340–355). doi: 10.1002/9780470960929.ch24

Hamdy, A. (2005). Saline irrigation management for a sustainable use. *Non-conventional water use: WASAMED project. CIHEAM/EU DG Research, Bari*, 3–42.

Han, Z. P., Liu, H., Zhao, X., Zhang, J., Guo, S., Wang, B., Zhao, L., Jin, Y., & Guo, Y. (2023). A maize phytochrome-interacting factor ZmPIF3 improves salt stress tolerance in transgenic Arabidopsis seedlings. *Research Square*. doi: 10.21203/rs.3.rs-3012134/v1

Hao, L., & Qiao, X. (2018). Genome-wide identification and analysis of the CNGC gene family in maize. *PeerJ*, 6(10), e5816. doi: 10.7717/peerj.5816

Haribabu, T. E., & Sudha, P. N. (2011). Effects of heavy metals copper and cadmium exposure on the antioxidants properties of the plant cleome gynandra. *Ijpaes*, 1(2), 80–87.

Hasan, M. I., Kibria, M. G., Jahiruddin, M., Murata, Y., & Hoque, M. A. (2015). Improvement of salt tolerance in maize by exogenous application of proline. *Journal of Environmental Science and Natural Resources*, 8(1), 13–18.

Hasan, M. K., Islam, M. S., Islam, M. R., Ismaan, H. N., Meena, R. S., & Barutcular, C. et al. (2019). Water relations and dry matter accumulation of black gram and mungbean as affected by salinity. *Thai Journal of Agricultural Science*, 52, 54–67.

Hasanuzzaman, M., Hossain, M. A., da Silva, J. A. T., & Fujita, M. (2012). Plant response and tolerance to abiotic oxidative stress: Antioxidant defense is a key factor. In: B. Venkateswarlu, A. Shanker, C. Shanker, & M. Maheswari, eds., *Crop stress and its management: Perspectives and strategies*. Springer. doi: 10.1007/978-94-007-2220-0_8

Hasanuzzaman, M., Nahar, K., & Fujit, M. (2013). Extreme temperature responses, oxidative stress and antioxidant defense in plants. In: K. Vahdati & C. Leslie, eds., *Abiotic stress—Plant responses and applications in agriculture* (pp. 169–205. doi: 10.5772/54833

Hasanuzzaman, M., Nahar, K., & Fujita, M. (2012). Plant response to salt stress and role of exogenous protectants to mitigate salt-induced damages. In: *Ecophysiology and responses of plants under salt stress* (pp. 25–87). New York, NY: Springer New York.

Hasegawa, P. M., Bressan, R. A., Zhu, J. K., & Bohnert, H. J. (2000). Plant cellular and molecular responses to high salinity. *Annual Review of Plant Biology*, 51, 463–499.

Hashem, A., Kumar, A., Al-Dbass, A. M., Alqarawi, A. A., Al-Arjani, A. B. F., & Singh, G. et al. (2019). Arbuscular mycorrhizal fungi and biochar improves drought tolerance in chickpea. *Saudi Journal of Biological Sciences*, 26, 614–624.

Hashimoto, K., & Kudla, J. (2011). Calcium decoding mechanisms in plants. *Biochimie*, 93, 2054–2059. doi: 10.1016/j.biochi.2011.05.019

Hatfield, J. L., & Prueger, J. H. (2015). Temperature extremes: Effect on plant growth and development. *Weather and Climate Extremes*, 10, 4–10. doi: 10.1016/j.wace.2015.08.001

Hawkins, E., Fricker, T. E., Challinor, A. J., Ferro, C. A. T., Ho, C. K., & Osborne, T. M. (2013). Increasing influence of heat stress on French maize yields from the 1960s to the 2030s. *Global Change Biology*, 19(3), 937–947. doi: 10.1111/gcb.12069

Hayes, S., Schachtschabel, J., Mishkind, M., Munnik, T., & Arisz, S. A. (2021). Hot topic: Thermosensing in plants. *Plant Cell and Environment*, 44, 2018–2033. doi: 10.1111/pce.13979

He, F. H., Pan, Y. H., Tan, L. L., Zhang, Z. H., Li, P., & Liu, J. et al. (2017). Study of the water transportation characteristics of marsh saline soil in the Yellow River delta. *Science of the Total Environment*, 574, 716–723.

He, R.-y., Wang, G.-j., & and Wang, X.-s. (1991). Effects of brassinolide on growth and chilling resistance of maize seedlings, 220–230.

He, Z., Zhong, J., Sun, X., Wang, B., Terzaghi, W., & Dai, M. (2018). The maize ABA receptors ZmPYL8, 9, and 12 facilitate plant drought resistance. *Frontiers in Plant Science*, 9, 422. doi: 10.3389/fpls.2018.00422

Hiyane, R., Hiyane, S., Tang, A. C., & Boyer, J. S. (2010). Sucrose feeding reverses shade-induced kernel losses in maize. *Annals of Botany*, 106, 395–403.

Hodges, D. M., Andrews, C. J., Johnson, D. A., & Hamilton, R. I. (1997a). Antioxidant enzyme and compound responses to chilling stress and their combining abilities in differentially sensitive maize hybrids. *Crop Science*, 37, 857–863. doi: 10.2135/cropsci1997.0011183X003700030027x

Hodges, D. M., Andrews, C. J., Johnson, D. A., & Hamilton, R. I. (1997b). Sensitivity of maize hybrids to chilling and their combining abilities at two developmental stages. *Crop Science*, 37, 850–856. doi: 10.2135/cropsci1997.0011183X003700030026x

Horváth, I., Glatz, A., Nakamoto, H., Mishkind, M. L., Munnik, T., Saidi, Y., Goloubinoff, P., Harwood, J. L., & Vigh, L. (2012). Heat shock response in photosynthetic organisms: Membrane and lipid connections. *Progress in Lipid Research*, 51, 208–220. doi: 10.1016/j.plipres.2012.02.002

Hossain, M. A., Hasanuzzaman, M., & Fujita, M. (2010). Up-regulation of antioxidant and glyoxalase systems by exogenous glycinebetaine and proline in mung bean confer tolerance to cadmium stress. *Physiology and Molecular Biology of Plants*, 16(3), 259–272.

Hossain, M. A., Piyatida, P., Silva, J. A. T., & Fujita, M. (2012). Molecular mechanism of heavy metal toxicity and tolerance in plants: Central role of glutathione in detoxification of reactive oxygen species and methylglyoxal and in heavy metal chelation. *Journal of Botany*, 2012, 1–37.

Hu, D., Li, R., Dong, S., Zhang, J., Zhao, B., & Ren, B. et al. (2022). Maize (*Zea mays* L.) responses to salt stress in terms of root anatomy, respiration and antioxidative enzyme activity. *BMC Plant Biology*, 22, 602. doi: 10.1186/s12870-022-03972-4

Hu, G., Li, Z., Lu, Y., Li, C., Gong, S., & Yan, S. et al. (2017). Genome-wide association study identified multiple genetic loci on chilling resistance during germination in maize. *Scientific Reports*, 7, 10840. doi: 10.1038/s41598-017-11318-6

Hu, X., Li, Y., Li, C., Yang, H., Wang, W., & Lu, M. (2010). Characterization of small heat shock proteins associated with maize tolerance to combined drought and heat stress. *Journal of Plant Growth Regulation*, 29, 455–464. doi: 10.1007/s00344-010-9157-9

Hu, X., Lu, M., Li, C., Liu, T., Wang, W., & Wu, J. et al. (2011). Differential expression of proteins in maize roots in response to abscisic acid and drought. *Acta Physiologiae Plantarum*, 33, 2437–2446. doi: 10.1007/s11738-011-0784-y

Hu, X., Wang, C., Fu, Y., Liu, Q., Jiao, X., & Wang, K. (2016). Expanding the range of CRISPR/Cas9 genome editing in rice. *Molecular Plant*, 9, 943–945. doi: 10.1016/j.molp.2016.03.003

Hu, X., Wang, G., Du, X., Zhang, H., Xu, Z., & Wang, J. et al. (2021). QTL analysis across multiple environments reveals promising chromosome regions associated with yield-related traits in maize under drought conditions. *The Crop Journal*, 9, 759–766. doi: 10.1016/j.cj.2020.10.004

Hu, X., Wu, X., Li, C., Lu, M., Liu, T., & Wang, Y. et al. (2012). Abscisic acid refines the synthesis of chloroplast proteins in maize (Zea mays) in response to drought and light. *PLoS One*, 7, e49500. doi: 10.1371/journal.pone.0049500

Huang, C. H., Xue, X., Wang, T., Mascellis, R. D., Mele, G., & You, Q. G. et al. (2011). Effects of saline water irrigation on soil properties in northwest China. *Environmental Earth Sciences*, 63, 701–708.

Huang, H., Møller, I. M., & Song, S. Q. (2012). Proteomics of desiccation tolerance during development and germination of maize embryos. *Journal of Proteomics*, 75, 1247–1262. doi: 10.1016/j.jprot.2011.10.036

Huang, J., Zhang, J., Li, W., Hu, W., Duan, L., & Feng, Y. et al. (2013). Genome-wide association analysis of ten chilling tolerance indices at the germination and seedling stages in maize. *Journal of Integrative Plant Biology*, 55, 735–744. doi: 10.1111/jipb.12051

Huang, M., Zhang, Z., Zhu, C., Zhai, Y., & Lu, P. (2019). Effect of biochar on sweet corn and soil salinity under conjunctive irrigation with brackish water in coastal saline soil. *Scientia Horticulturae*, 250, 405–413.

Huang, Y.-C., Niu, C.-Y., Yang, C.-R., & Jinn, T.-L. (2016). The heat stress factor HSFA6b connects ABA signaling and ABA-mediated heat responses. *Plant Physiology*, 172(2), 1182–1199. doi: 10.1104/pp.16.00860

Hund, A., Fracheboud, Y., Soldati, A., & Stamp, P. (2008). Cold tolerance of maize seedlings as determined by root morphology and photosynthetic traits. *European Journal of Agronomy*, 28, 178–185. doi: 10.1016/j.eja.2007.07.003

Hussain, H. A., Hussain, S., Khaliq, A., Ashraf, U., Anjum, S. A., & Men, S. et al. (2018). Chilling and drought stresses in crop plants: Implications, cross talk, and potential management opportunities. *Frontiers in Plant Science*, 9, 393. doi: 10.3389/fpls.2018.00393

Hussain, S. S., Rasheed, M., Ahmed, Z. I., & Jilani, G. (2023). Characterizing maize genotypes for salt tolerance using morphological and ionic traits at seedling stage. *International Letters of Natural Sciences*, 86, 1–24.

Hussain, Z., Khattak, R. A., Fareed, I., Irshad, M., & Mahmood, Q. (2015). Interaction of phosphorus and potassium on maize (*Zea mays* L.) in saline-sodic soil. *Journal of Agricultural Science*, 7, 66–78.

Inghelandt, D. V., Frey, F. P., Ries, D., & Stich, B. (2019). QTL mapping and genome-wide prediction of heat tolerance in multiple connected populations of temperate maize. *Scientific Reports*, 9(1), 14418. doi: 10.1038/s41598-019-50853-2

IPCC, Fourth Assessment Report: Synthesis, published online 17 November 2007.

Iqbal, H., Yaning, C., Rehman, H. U., Waqas, M., Ahmed, Z., Raza, S. T., & Shareef, M. (2020). Improving heat stress tolerance in late planted spring maize by using different exogenous elicitors. *Chilean Journal of Agricultural Research*, 80(1), 30–40. doi: 10.4067/S0718-58392020000100030

Iqbal, M., Ashraf, M., & Jamil, A. (2006). Seed enhancement with cytokinins: Changes in growth and grain yield in salt stressed wheat plants. *Plant Growth Regulation*, 50, 29–39.

Islam, M. S., Akhter, M. M., El Sabagh, A., Liu, L. Y., Nguyen, N. T., & Ueda, A. et al. (2011). Comparative studies on growth and physiological responses to saline and alkaline stresses of foxtail millet (*Setaria italica* L.) and Proso millet (*Panicum miliaceum* L.). *Australian Journal of Crop Science*, 5, 1269–1277.

Islam, M.S., Sabagh, A.E., Hasan, K., Akhter, M., Barutçular, C. (2017a). Growth and yield response of mungbean (Vignaradiata L.) as influenced by sulphur and boron application. *Scientific Journal of Crop Science*, 6(1): 153–160.

Islam, M. S., Hasan, M. K., Islam, M. R., Chowdhury, M. K., Pramanik, M. H., & Iqbal, M. A. et al. (2023). Water relations and yield characteristics of mungbean as influenced by foliar application of gibberellic acid (GA3). *Frontiers in Ecology and Evolution*, 11, 1048768.

Ismail, M. A. (2013). Alleviation of salinity stress in white corn (*Zea mays* L.) plant by exogenous application of salicylic acid. *American Journal of Life Sciences*, 1, 248–255.

Janmohammadi, M., Dezfuli, P. M., & Sharifzadeh, F. (2008). Seed invigoration techniques to improve germination and early growth of inbred line of maize under salinity and drought stress. *General and Applied Plant Physiology*, 34, 215–226.

Janowiak, F., Maas, B., & Dörffling, K. (2002). Importance of abscisic acid for chilling tolerance of maize seedlings. *Journal of Plant Physiology*, 159, 635–643. doi: 10.1078/0176-1617-0638

Jha, U. C., Nayyar, H., Palakurthi, R., Jha, R., Valluri, V., Bajaj, P., ... & Thudi, M. (2021). Major QTLs and potential candidate genes for heat stress tolerance identified in chickpea (Cicer arietinum L.). *Frontiers in Plant Science*, 12, 655103.

Jia, T. J., Jing-Jing, L. I., Wang, L. F., Cao, Y. Y., Juan, M. A., & Hao, W. A. N. G. et al. (2020). Evaluation of drought tolerance in ZmVPP1-overexpressing transgenic inbred maize lines and their hybrids. *Journal of Integrative Agriculture*, 19, 2177–2187. doi: 10.1016/s2095-3119(19)62828-5

Jia, Z., Lian, Y., Zhu, Y., He, J., Cao, Z., & Wang, G. (2009). Cloning and characterization of a putative transcription factor induced by abiotic stress in *Zea mays*. *African Journal of Biotechnology*, 8(24), 6764–6771.

Jiang, C., Belfield, E. J., Cao, Y., Smith, J. A., & Harberd, N. P. (2013). An Arabidopsis soil-salinity-tolerance mutation confers ethylene-mediated enhancement of sodium/potassium homeostasis. *Plant Cell*, 25, 3535–3552.

Jin, Y., Zhang, Z., Xi, Y., Yang, Z., Xiao, Z., & Guan, S. et al. (2021). Identification and functional verification of cold tolerance genes in spring maize seedlings based on a genome-wide association study and quantitative trait locus mapping. *Frontiers in Plant Science*, 12, 776972. doi: 10.3389/fpls.2021.776972

Jonak, C., Okresz, L., Bogre, L., & Hirt, H. (2002). Complexity, cross talk and integration of plant MAP kinase signalling. *Current Opinion in Plant Biology*, 5, 415–424.

Joshi, J., Hasnain, G., Logue, T., Lynch, M., Wu, S., Guan, J.-C., Alseekh, S., Fernie, A. R., Hanson, A. D., & Mccarty, D. R. (2021). A core metabolome response of maize leaves subjected to long-duration abiotic stresses. *Metabolites*, 11(11), 797.

Juurakko, C. L., DiCenzo, G. C., & Walker, V. K. (2021). Cold acclimation and prospects for cold-resilient crops. *Plant Stress*, 2, 100028. doi: 10.1016/j.stress.2021.100028

Kadota, Y., Watanabe, T., Fujii, S., Maeda, Y., Ohno, R., & Higashi, K. et al. (2005). Cell cycle dependence of elicitor-induced signal transduction in tobacco BY-2 cells. *Plant and Cell Physiology*, 46, 156–165. doi: 10.1093/pcp/pci008

Kamrul, H. M., Islam, M. S., Islam, M. R., Ismaan, H. N., & El-Sabagh, A. (2018). Germination and early seedling growth of mungbean (*Vigna radiata* L.) as influenced by salinity. *Azarian Journal of Agriculture*, 5, 49–59.

Kaplan, B., Sherman, T., & Fromm, H. (2007). Cyclic nucleotide-gated channels in plants. *FEBS Letters*, 581(12), 2237–2246. doi: 10.1016/j.febslet.2007.02.017

Karim, M. A., Frach eboud, Y., & Stamp, P. (1999). Photosynthetic activity of developing leaves of Zea mays is less affected by heat stress than that of developed leaves. *Physiologia Plantarum*, 105, 685–693. doi: 10.1034/j.1399-3054.1999.105413.x

Karimi, G., Ghorbanli, M., Heidari, H., Khavari Nejad, R. A., & Assareh, M. H. (2005). The effects of NaCl on growth, water relations, osmolytes and ion content in *Kochia prostrata*. *Biologia Plantarum*, 49, 301–304.

Kartal, G., Temelm, A., Aricanm, E., & Gozukirmizim, N. (2009). Effects of brassinosteroids on barley root growth, antioxidant system and cell division. *Plant Growth Regulation*, 58, 261–267.

Kaya, C., Tuna, A. L., & Okant, A. M. (2010). Effect of foliar applied kinetin and indoleacetic acid on maize plants grown undersaline conditions. *Turkish Journal of Agriculture and Forestry*, 34, 529–538.

Kaya, M., Kaya, G., Kaya, M. D., Atak, M., Saglam, S., & Khawar, K. M. et al. (2008). Interaction between seed size and NaCl on germination and early seedling growth of some Turkish cultivars of chickpea (*Cicer arietinum* L.). *Journal of Zhejiang University Science B*, 9, 371–377.

Kebede, H., Fisher, D. K., & Young, L. D. (2012). Determination of moisture deficit and heat stress tolerance in corn using physiological measurements and a low-cost microcontroller-based monitoring system. *Journal of Agronomy and Crop Science*, 198(2), 118–129. doi: 10.1111/j.1439-037X.2011.00493.x

Keskin, B. C., Yuksel, B., Memon, A. R., & Topal-Sarıkaya, A. (2010). Abscisic acid regulated gene expression in bread wheat ('Triticum aestivum' L.). *Australian Journal of Crop Science*, 4(8), 617–625.

Kettler, B. A., Carrera, C. S., Sonzogni, F. D. N., Trachsel, S., Andrade, F. H., & Neiff, N. (2022). High night temperature during maize post flowering increases night respiration and reduces photosynthesis, growth, and kernel number. *Journal of Agronomy and Crop Science*, 208(3), 335–347. doi: 10.1111/jac.12589

Khajeh-Hosseini, M., Powell, A. A., & Bingham, I. J. (2003). The interaction between salinity stress and seed vigour during germination of soyabean seeds. *Seed Science and Technology*, 31, 715–725.

Khalil, A. M. (2020). The genome editing revolution. *Journal of Genetic Engineering and Biotechnology*, 18, 1–16. doi: 10.1093/oso/9780198766834.003.0001

Khan, S., Anwar, S., Yu, S., Sun, M., Yang, Z., & Gao, Z. Q. (2019). Development of drought-tolerant transgenic wheat: achievements and limitations. *International Journal of Molecular Sciences*, 20, 3350. doi: 10.3390/ijms20133350

Khan, A., Khan, A. L., Muneer, S., Kim, Y. H., Al-Rawahi, A., & Al-Harrasi, A. (2019). Silicon and salinity: Crosstalk in crop-mediated stress tolerance mechanisms. *Frontiers in Plant Science*, 10, 1429.

Khan, W. U., Aziz, T., Maqsood, M. A., Farooq, M., Abdullah, Y., & Ramzani, P. M. et al. (2018). Silicon nutrition mitigates salinity stress in maize by modulating ion accumulation, photosynthesis, and antioxidants. *Photosynthetica*, 56, 1047–1057.

Khān, M. A., & Weber, D. J. (2008). *Ecophysiology of high salinity tolerant plants*. Amsterdam, Netherlands: Springer.

Khodarahmpour, Z., Ifar, M., & Motamedi, M. (2012). Effects of NaCl salinity on maize (*Zea mays* L.) at germination and early seedling stage. *African Journal of Biotechnology*, 11, 298–304.

Khodary, S. E. A. (2004). Effect of salicylic acid on the growth, photosynthesis and carbohydrate metabolism in salt-stressed maize plants. *International Journal of Agriculture and Biology*, 6, 5–8.

Kikoyo, D. A., & Nobert, J. (2016). Assessment of the impact of climate change on runoff and crop yield using SWAT model: A case of Kibungo sub-catchment, Uganda. *Physics and Chemistry of the Earth*, 93, 96–103.

Kimotho, R. N., Baillo, E. H., & Zhang, Z. (2019). Transcription factors involved in abiotic stress responses in maize (Zea mays L.) and their roles in enhanced productivity in the post genomics era. *PeerJ*, 7, e7211. doi: 10.7717/peerj.7211

Kong, X., Pan, J., Zhang, D., Jiang, S., Cai, G., & Wang, L. et al. (2013). Identification of mitogen-activated protein kinase gene family and MKK-MAPK interaction network in maize. *Biochemical and Biophysical Research Communications*, 441, 964–969. doi: 10.1016/j.bbrc.2013.11.008

Kong, X., Pan, J., Zhang, M., Xing, X., Zhou, Y., & Liu, Y. et al. (2011). ZmMKK4, a novel group C mitogen-activated protein kinase in maize (Zea mays), confers salt and cold tolerance in transgenic Arabidopsis. *Plant, Cell & Environment*, 34, 1291–1303. doi: 10.1111/j.1365-3040.2011.02329.x

Konuşkan, Ö, Gözübenli, H., Atış, İ, & Atak, M. (2017). Effects of salinity stress on emergence and seedling growth parameters of some maize genotypes (*Zea mays* L.). *Turkish Journal of Agriculture - Food Science and Technology*, 5, 1668–1672.

Kotak, S., Port, M., Ganguli, A., Bicker, F., & Von Koskull-Döring, P. (2004). Characterization of C-terminal domains of Arabidopsis heat stress transcription factors (Hsfs) and identification of a new signature combination of plant class a Hsfs with AHA and NES motifs essential for activator function and intracellular localization. *Plant Journal*, 39(1), 98–112. doi: 10.1111/j.1365-313X.2004.02111.x

Krzesłowska, M. (2011). The cell wall in plant cell response to trace metals: Polysaccharide remodeling and its role in defense strategy. *Acta Physiologiae Plantarum*, 33(1), 35–51.

Kumar, A., Carter, A., Mir, R. R., Sehgal, D., Agarwal, P., & Paterson, A. H. (2022). Genetics and genomics to enhance crop production, towards food security. *Frontiers in Genetics*, 12, 798308. doi: 10.3389/fgene.2021.798308

Kumar, A., Dames, J. F., Gupta, A., Sharma, S., Gilbert, J. A., & Ahmad, P. (2015). Current developments in arbuscular mycorrhizal fungi research and its role in salinity stress alleviation: A biotechnological perspective. *Critical Reviews in Biotechnology*, 35, 461–474.

Kumar, A., Sharma, S., & Mishra, S. (2016). Evaluating effect of arbuscular mycorrhizal fungal consortia and azotobacter chroococcum in improving biomass yield of *Jatropha curcas*. *Plant Biosystems*, 150, 1056–1064.

Lamers, J., van der Meer, T., & Testerink, C. (2020). How plants sense and respond to stressful environments. *Plant Physiology*, 182, 1624–1635. doi: 10.1104/pp.19.01464

Langridge, P., & Fleury, D. (2011). Making the most of 'omics' for crop breeding. *Trends in Biotechnology*, 29, 33–40.

Lata, C., Muthamilarasan, M., & Prasad, M. (2015). Drought stress responses and signal transduction in plants. In: G. K. Pandey, ed. *Elucidation of abiotic stress signaling in plants (pp.* 195–225). New York, NY: Springer. doi: 10.1007/978-1-4939-2540-7_7

Li, B., Fan, R., Sun, G., Sun, T., Fan, Y., Bai, S., Guo, S., Huang, S., Liu, J., Zhang, H., Wang, P., Zhu, X., & Song, C.-P. (2021). Flavonoids improve drought tolerance of maize seedlings by regulating the homeostasis of reactive oxygen species. *Plant and Soil*, 461(1), 389–405. doi: 10.1007/s11104-020-04814-8

Li, B., & Fang, Z. (2018). Screening of low-temperature tolerance maize inbred lines and analysis of leaf physiological characteristics and cellular structure changes. *Journal of Henan Agricultural Sciences*, 47, 31–37.

Li, H., Han, X., Liu, X., Zhou, M., Ren, W., & Zhao, B. et al. (2019). A leucine-rich repeat-receptor-like kinase gene SbER2–1 from sorghum (Sorghum bicolor L.) confers drought tolerance in maize. *BMC Genomics*, 20, 737. doi: 10.1186/s12864-019-6143-x

Li, H., Yue, H., Xie, J., Bu, J., Li, L., & Xin, X. et al. (2021). Transcriptomic profiling of the high-vigour maize (Zea mays L.) hybrid variety response to cold and drought stresses during seed germination. *Scientific Reports*, 11, 1–16. doi: 10.1038/s41598-021-98907-8

Li, H. Y., Huang, S. H., Shi, Y. S., Song, Y. C., Zhong, Z. B., & Wang, G. Y. et al. (2009). Isolation and analysis of drought-induced genes in maize roots. *Agricultural Sciences in China*, 8, 129–136. doi: 10.1016/s1671-2927(09)60019-5

Li, J., Zhu, Q., Jiao, F., Yan, Z., Zhang, H., & Zhang, Y. et al. (2023). Research progress on the mechanism of salt tolerance in maize: A classic field that needs new efforts. *Plants*, 12, 2356.

Li, J. F., Norville, J. E., Aach, J., McCormack, M., Zhang, D., & Bush, J. et al. (2013). Multiplex and homologous recombination-mediated genome editing in Arabidopsis and Nicotianabenthamiana using guide RNA and Cas9. *Nature Biotechnology*, 31, 688–691. doi: 10.1038/nbt.2654

Li, M., Sui, N., Lin, L., Yang, Z., & Zhang, Y. (2019b). Transcriptomic profiling revealed genes involved in response to cold stress in maize. *Functional Plant Biology*, 46, 830–844. doi: 10.1071/FP19065

Li, Z., & Howell, S. H. (2021). Heat stress responses and thermotolerance in maize. *International Journal of Molecular Sciences*, 22(2), 948. doi: 10.3390/ijms22020948

Li, Z., Hu, G., Liu, X., Zhou, Y., Li, Y., & Zhang, X. et al. (2016). Transcriptome sequencing identified genes and gene ontologies associated with early freezing tolerance in maize. *Frontiers in Plant Science*, 7, 1477. doi: 10.3389/fpls.2016.01477

Li, Z., Xu, J., Gao, Y., Wang, C., Guo, G., & Luo, Y. et al. (2017). The synergistic priming effect of exogenous salicylic acid and H2O2 on chilling tolerance enhancement during maize (Zea mays L.) seed germination. *Frontiers in Plant Science*, 8, 1153. doi: 10.3389/fpls.2017.01153

Li, Y., Wang, X., Li, Y., Zhang, Y., Gou, Z., Qi, X., & Zhang, J. (2020). Transcriptomic analysis revealed the common and divergent responses of maize seedling leaves to cold and heat stresses. *Genes (Basel)*, 11(8), 881. doi: 10.3390/genes11080881

Lin, Y.-X., Jiang, H.-Y., Chu, Z.-X., Tang, X.-L., Zhu, S.-W., & Cheng, B.-J. (2011). Genome-wide identification, classification and analysis of heat shock transcription factor family in maize. *BMC Genomics*, 12, 76. doi: 10.1186/1471-2164-12-76

Lipka, A. E., Kandianis, C. B., Hudson, M. E., Yu, J., & Drnevich, J. et al. (2015). From association to prediction: Statistical methods for the dissection and selection of complex traits in plants. *Current Opinion in Plant Biology*, 24, 110–118. doi: 10.1016/j.pbi.2015.02.010

Liu, J., Niu, Y., Zhang, J., Zhou, Y., Ma, Z., & Huang, X. (2018). Ca^{2+} channels and Ca^{2+} signals involved in abiotic stress responses in plant cells: Recent advances. *Plant Cell, Tissue and Organ Culture*, 132, 413–424. doi: 10.1007/s11240-017-1350-0

Liu, S., Wang, X., Wang, H., Xin, H., Yang, X., & Yan, J. et al. (2013). Genome-wide analysis of ZmDREB genes and their association with natural variation in drought tolerance at seedling stage of Zea mays L. *PLoS Genetics*, 9, e1003790. doi: 10.1371/journal.pgen.1003790

Liu, Y., Qin, L., Han, L., Xiang, Y., & Zhao, D. (2015). Overexpression of maize SDD1 (ZmSDD1) improves drought resistance in Zea mays L. by reducing stomatal density. *Plant Cell, Tissue and Organ Culture*, 122, 147–159. doi: 10.1007/s11240-015-0757-8

Liu, Y., Subhash, C., Yan, J., Song, C., Zhao, J., & Li, J. (2011). Maize leaf temperature responses to drought: Thermal imaging and quantitative trait loci (QTL) mapping. *Environmental and Experimental Botany*, 71, 158–165. doi: 10.1016/j.envexpbot.2010.11.010

Lizaso, J. I., Ruiz-Ramos, M., Rodríguez, L., Gabaldon-Leal, C., Oliveira, J. A., Lorite, I. J., Sánchez, D., García, E., & Rodríguez, A. (2018). Impact of high temperatures in maize: Phenology and yield components. *Field Crops Research*, 216, 129–140. doi: 10.1016/j.fcr.2017.11.013

Lobell, D. B., Bänziger, M., Magorokosho, C., & Vivek, B. (2011). Nonlinear heat effects on African maize as evidenced by historical yield trials. *Nature Climate Change*, 1, 42–45.

Lobell, D. B., & Burke, M. B. (2010). On the use of statistical models to predict crop yield responses to climate change. *Agricultural and Forest Meteorology*, 150, 1443–1452.

Lobell, D. B., Burke, M. B., Tebaldi, C., Mastrandrea, M. D., & Falcon, W. P. et al. (2008). Prioritizing climate change adaptation needs for food security in 2030. *Science*, 319(5863), 607–610.

Lobell, D. B., Schlenker, W., & Costa-Roberts, J. (2011). Climate trends and global crop production since 1980. *Science*, *333*(6042), 616–620. doi: 10.1126/science.1204531

Lohani, N., Singh, M. B., & Bhalla, P. L. (2020). High temperature susceptibility of sexual reproduction in crop plants. *Journal of Experimental Botany*, 71(2), 555–568. doi: 10.1093/jxb/erz426

Lohaus, G., Hussmann, M., Pennewiss, K., Schneider, H., Zhu, J. J., & Sattelmacher, B. (2000). Solute balance of a maize (*Zea mays* L.) source leaf as affected by salt treatment with special emphasis on phloem retranslocation and ion leaching. *Journal of Experimental Botany*, 51, 1721–1732.

Lorenz, A. J. (2013). Resource allocation for maximizing prediction accuracy and genetic gain of genomic selection in plant breeding: A simulation experiment. *G3 Genes Genomes Genetics*, 3, 481–491. doi: 10.1534/g3.112.004911

Lu, G. H., Tang, J. H., Yan, J. B., Ma, X. Q., Li, J. S., & Chen, S. J. et al. (2006). Quantitative trait loci mapping of maize yield and its components under different water treatments at flowering time. *Journal of Integrative Plant Biology*, 48, 1233–1243. doi: 10.1111/j.1744-7909.2006.00289.x

Lu, X., Yang, L., Yu, M., Lai, J., Wang, C., & McNeil, D. et al. (2017). A novel Zea mays ssp. mexicana L.MYC-type ICE-like transcription factor gene ZmmICE1, enhances freezing tolerance in transgenic Arabidopsis thaliana. *Plant Physiology and Biochemistry*, 113, 78–88. doi: 10.1016/j.plaphy.2017.02.002

Lu, Y., Hao, Z., Xie, C., Crossa, J., Araus, J. L., & Gao, S. et al. (2011). Large-scale screening for maize well-watered environments. *Field Crop Research*, 124, 37–45. doi: 10.1016/j.fcr.2011.06.003

Lu, Y., Li, Y., Zhang, J., Xiao, Y., Yue, Y., & Duan, L. et al. (2013). Overexpression of Arabidopsis molybdenum cofactor sulfurase gene confers drought tolerance in maize (Zea mays L.). *PLoS One*, 8, e52126. doi: 10.1371/journal.pone.0052126

Luna, E., Pastor, V., Robert, J., Flors, V., Mauch-Mani, B., & Ton, J. (2011). Callose deposition: A multifaceted plant defense response. *Molecular Plant-Microbe Interactions*, 24, 183–193. doi: 10.1094/MPMI-07-10-0149

Luo, M., Liu, J., Lee, R. D., Scully, B. T., & Guo, B. (2010). Monitoring the expression of maize genes in developing kernels under drought stress using oligo-microarray. *Journal of Integrative Plant Biology*, 52, 1059–1074. doi: 10.1111/j.1744-7909.2010.01000.x

Luo, M., Zhang, Y., Li, J., Zhang, P., Chen, K., & Song, W. et al. (2021). Molecular dissection of maize seedling salt tolerance using a genome-wide association analysis method. *Plant Biotechnology Journal*, 19, 1937–1951.

Luo, X., Wang, B., Gao, S., Zhang, F., Terzaghi, W., & Dai, M. (2019). Genome-wide association study dissects the genetic bases of salt tolerance in maize seedlings. *Journal of Integrative Plant Biology*, 61, 658–674.

Lyon, D., Castillejo, M. A., Mehmeti-Tershani, V., Staudinger, C., Kleemaier, C., & Wienkoop, S. (2016). Drought and recovery: Independently regulated processes highlighting the importance of protein turnover dynamics and translational regulation in Medicago truncatula. *Molecular & Cellular Proteomics*, 15, 1921–1937.

Ma, J., Xing, G., Yang, X., Wang, Y., & Du, H. (2015a). Effects of exogenous EBR and NO signal on antioxidant system and low response gene expression under cold stress on maize embryo. *Chinese Journal of Applied Ecology*, 26, 1411–1418.

Maazou, A. R. S., Tu, J., Qiu, J., & Liu, Z. (2016). Breeding for drought tolerance in maize (Zea mays L.). *American Journal of Plant Sciences*, 7, 1858.

Madhumal Thayil, V., Zaidi, P. H., Seetharam, K., Rani Das, R., Viswanadh, S., Ahmed, S., Miah, M. A., Koirala, K. B., Tripathi, M. P., & Arshad, M. et al. (2020). Genotype-by-environment interaction effects under heat stress in tropical maize. *Agronomy*, 10(12), 1998. doi: 10.3390/agronomy10121998

Malash, N., Flowers, T. J., & Ragab, R. (2005). Effect of irrigation systems and water management practices using saline and non-saline water on tomato production. *Agricultural Water Management*, 78, 25–38.

Malenica, N., Dunić, J. A., Vukadinović, L., Cesar, V., & Šimić, D. (2021). Genetic approaches to enhance multiple stress tolerance in maize. *Genes*, 12, 1760. doi: 10.3390/genes12111760

Mao, H., Wang, H., Liu, S., Li, Z., Yang, X., & Yan, J. et al. (2015). Qin A transposable element in a NAC gene is associated with drought tolerance in maize seedlings. *Nature Communications*, 6, 8326. doi: 10.1038/ncomms9326

Maqsood, T., Akhtar, J., Farooq, M. R., Haq, M. A., & Saqib, Z. A. (2008). Biochemical attributes of salt tolerant and salt sensitive maize cultivars to salinity and potassium nutrition. *Pakistan Journal of Agricultural Sciences*, 45, 1–5.

Martignago, D., Rico-Medina, A., Blasco-Escámez, D., Fontanet-Manzaneque, J. B., & Caño-Delgado, A. I. (2020). Drought resistance by engineering plant tissue-specific responses. *Frontiers in Plant Science*, 10, 1676. doi: 10.3389/fpls.2019.01676

Masle, J., Gilmore, S. R., & Farquhar, G. D. (2005). The ERECTA gene regulates plant transpiration efficiency in Arabidopsis. *Nature*, 436, 866–870. doi: 10.1038/nature03835

Matika, D. E. F., & Loake, G. J. (2014). Redox regulation in plant immune function. *Antioxidants & Redox Signaling*, 21, 1373–1388. doi: 10.1089/ars.2013.5679

McNellie, J. P., Chen, J., Li, X., & Yu, J. (2018). Genetic mapping of foliar and tassel heat stress tolerance in maize. *Crop Science*, 58(6), 2484–2493. doi: 10.2135/cropsci2018.05.0291

Mei, X., Zhao, Z., Bai, Y., Yang, Q., Gan, Y., & Wang, W. et al. (2023). Salt tolerant Gene 1 contributes to salt tolerance by maintaining photosystem II activity in maize. *Plant, Cell & Environment*, 46, 1833–1848.

Meloni, D. A., Oliva, M. A., Ruiz, H. A., & Martinez, C. A. (2001). Contribution of proline and inorganic solutes to osmotic adjustment in cotton under salt stress. *Journal of Plant Nutrition*, 24, 599–612.

Meng, C., & Sui, N. (2019). Overexpression of maize MYB-IF35 increases chilling tolerance in Arabidopsis. *Plant Physiology and Biochemistry*, 135, 167–173. doi: 10.1016/j.plaphy.2018.11.038

Mibulo, M., & Kiggundu, N. (2018). Impact of water harvesting techniques on maize yield using AquaCrop. *Agricultural Sciences*, 9, 1434–1451.

Minhas, P. S. (1996). Saline water management for irrigation in India. *Agricultural Water Management*, 30, 1–24.

Mittler, R., Finka, A., & Goloubinoff, P. (2012). How do plants feel the heat? *Trends in Biochemical Sciences*, 37, 118–125. doi: 10.1016/j.tibs.2011.11.007

Momčilović, I., & Ristic, Z. (2007). Expression of chloroplast protein synthesis elongation factor, EF-Tu, in two lines of maize with contrasting tolerance to heat stress during early stages of plant development. *Journal of Plant Physiology*, 164(1), 90–99. doi: 10.1016/j.jplph.2006.01.010

Moradtalab, N., Weinmann, M., Walker, F., Höglinger, B., Ludewig, U., & Neumann, G. (2018). Silicon improves chilling tolerance during early growth of maize by effects on micronutrient homeostasis and hormonal balances. *Frontiers in Plant Science*, 9, 420. doi: 10.3389/fpls.2018.00420

Munns, R. (2002). Comparative physiology of salt and water stress. *Plant, Cell & Environment*, 25(2), 239–250.

Munns, R., James, R. A., & Läuchli, A. (2006). Approaches to increasing the salt tolerance of wheat and other cereals. *Journal of Experimental Botany*, 57, 1025–1043.

Munns, R., & Tester, M. (2008). Mechanisms of salinity tolerance. *Annual Review of Plant Biology*, 59, 651–681.

Muppala, S., Gudlavalleti, P. K., Malireddy, K. R., Puligundla, S. K., & Dasari, P. (2021). Development of stable transgenic maize plants tolerant for drought by manipulating ABA signaling through agrobacterium-mediated transformation. *Journal of Genetic Engineering and Biotechnology*, 19, 1–14. doi: 10.1186/s43141-021-00195-2

Murillo-Amador, B., López-Aguilar, R., Kaya, C., Larrinaga-Mayoral, J., & Flores-Hernández, A. (2002). Comparative effects of NaCl and polyethylene glycol on germination, emergence and seedling growth of cowpea. *Journal of Agronomy and Crop Science*, 188, 235–247.

Mwamahonje, A., Eleblu, J. S. Y., Ofori, K., Deshpande, S., Feyissa, T., & Tongoona, P. (2021). Drought tolerance and application of marker-assisted selection in sorghum. *Biology*, 10, 1249. doi: 10.3390/biology10121249

Nadarajah, K. K. (2020). ROS homeostasis in abiotic stress tolerance in plants. *International Journal of Molecular Science*, 21, 5208. doi: 10.3390/ijms21155208

Nadeem, M., Li, J., Yahya, M., Sher, A., Ma, C., & Wang, X. et al. (2019). Research progress and perspective on drought stress in legumes: A review. *International Journal of Molecular Science*, 20, 2541. doi: 10.3390/ijms20102541

Neill, E. M., Byrd, M. C. R., Billman, T., Brandizzi, F., & Stapleton, A. E. (2019). Plant growth regulators interact with elevated temperature to alter heat stress signaling via the unfolded protein response in maize. *Scientific Reports*, 9(1), 10392. doi: 10.1038/s41598-019-46839-9

Nelimor, C., Badu-Apraku, B., Tetteh, A. Y., & N'guetta, A. S. (2019). Assessment of genetic diversity for drought, heat and combined drought and heat stress tolerance in early maturing maize landraces. *Plants*, 8(11), 518.

Nelson, D. E., Repetti, P. P., Adams, T. R., Creelman, R. A., Wu, J., & Warner, D. C. et al. (2007). Plant nuclear factor Y (NF-Y) B subunits confer drought tolerance and lead to improved corn yields on water-limited acres. *Proceedings of the National Academy of Sciences of the United States of America*, 104, 16450–16455. doi: 10.1073/pnas.0707193104

Nemali, K. S., Bonin, C., Dohleman, F. G., Stephens, M., Reeves, W. R., & Nelson, D. E. et al (2015). Physiological responses related to increased grain yield under drought in the first biotechnology-derived drought-tolerant maize. *Plant, Cell & Environment*, 38, 1866–1880. doi: 10.1111/pce.12446

Neta, I. C. S., Von Pinho, ÉV. R., de Abreu, V. M., Vilela, D. R., Santos, M. C., & Dos Santos, H. O. et al. (2020). Gene expression and genetic control to cold tolerance during maize seed germination. *BMC Plant Biology*, 20, 188. doi: 10.1186/s12870-020-02387-3

Nikolić, A., Angelković, V., Dodig, D., & Ignjatović-Micić, D. (2011). Quantitative trait loci for yield and morphological traits in maize under drought stress. *Genetika*, 43, 263–276. doi: 10.2298/gensr1102263n

Niu, G., Xu, W., Rodriguez, D., & Sun, Y. (2012). Growth and physiological responses of maize and sorghum genotypes to salt stress. *ISRN Agronomy*, 2012, 145072.

Niu, S., Du, X., Wei, D., Liu, S., Tang, Q., Bian, D., Zhang, Y., Cui, Y., & Gao, Z. (2021). Heat stress after pollination reduces kernel number in maize by insufficient assimilates. *Frontiers in Genetics*, *12*, 728166. doi: 10.3389/fgene.2021.728166

Nocito, F. F., Lancilli, C., Crema, B., Fourcroy, P., Davidian, J.-C., & Sacchi, G. A. (2006). Heavy metal stress and sulfate uptake in maize roots. *Plant Physiology*, 141(3), 1138–1148. doi: 10.1104/pp.105.076240

Noor, J. J., Vinayan, M. T., Umar, S., Devi, P., Iqbal, M., Seetharam, K., & Zaidi, P. H. (2019a). Morpho-physiological traits associated with heat stress tolerance in tropical maize (*Zea mays* L.) at the reproductive stage. *Australian Journal of Crop Science*, 13(4), 536–545. doi: 10.21475/ajcs.19.13.04.p1448

Ohama, N., Sato, H., Shinozaki, K., & Yamaguchi-Shinozaki, K. (2017). Transcriptional regulatory network of plant heat stress response. *Trends in Plant Science*, 22, 53–65. doi: 10.1016/j.tplants.2016.08.015

Oladosu, Y., Rafii, M. Y., Samuel, C., Fatai, A., Magaji, U., & Kareem, I. et al. (2019). Drought resistance in rice from conventional to molecular breeding: A review. *International Journal of Molecular Science*, 20, 3519. doi: 10.3390/ijms20143519

Ouzounidou, G., Čiamporová, M., Moustakas, M., & Karataglis, S. (1995). Responses of maize (Zea mays L.) plants to copper stress—I. Growth, mineral content and ultrastructure of roots. *Environmental and Experimental Botany*, 35(2), 167–176.

Ovečka, M., & Takáč, T. (2014). Managing heavy metal toxicity stress in plants: Biological and biotechnological tools. *Biotechnology Advances*, 32(1), 73–86.

Ozturk, M., Altay, V., & Karahan, F. (2017). Studies on trace elements in *Glycyrrhiza* taxa distributed in Hatay-Turkey. *International Journal of Plant and Environment*, 3(2), 01–07.

Ozturk, M., Ashraf, M., Aksoy, A., & Ahmad, M. S. A., eds. (2015b). *Plants, pollutants & remediation*. New York: Springer.

Ozturk, M., Yucel, E., Gucel, S., Sakçali, S., & Aksoy, A. (2008). 28 Plants as Biomonitors of Trace Elements Pollution in Soil.

Pan, J., Zhang, M., Kong, X., Xing, X., Liu, Y., & Zhou, Y. et al. (2012). ZmMPK17, a novel maize group DMAP kinase gene, is involved in multiple stress responses. *Planta*, 235, 661–676. doi: 10.1007/s00425-011-1510-0

Parmar, N., Singh, K. H., Sharma, D., Singh, L., Kumar, P., & Nanjundan, J. et al. (2017). Genetic engineering strategies for biotic and abiotic stress tolerance and quality enhancement in horticultural crops: A comprehensive review. *Biotechnology*, 7, 1–35. doi: 10.1007/s13205-017-0870-y

Parrado, J. D., Canteros, F. H., & Lorea, R. (2021). Heat stress in maize: Characterization and phenotypic plasticity. *Maydica*, 65(3), 11–16.

Peng, Y., Wang, Y., Zhao, X., & Lv, Y. (2016). Evaluation of chilling tolerance in different maize inbred lines. *Agricultural Research in the Arid Areas*, 34, 267–280.

Pinhero, R. G., Rao, M. V., Paliyath, G., Murr, D. P., & Fletcher, R. A. (1997). Changes in activities of antioxidant enzymes and their relationship to genetic and paclobutrazol-induced chilling tolerance of maize seedlings. *Plant Physiology*, 114, 695–704. doi: 10.1104/pp.114.2.695

Porcel, R., Aroca, R., & Ruiz-Lozano, J. M. (2012). Salinity stress alleviation using arbuscular mycorrhizal fungi. A review. *Agronomy for Sustainable Development*, 32, 181–200.

Prasad, P. V. V., Bheemanahalli, R., & Jagadish, S. V. K. (2017). Field crops and the fear of heat stress – Opportunities, challenges and future directions. *Field Crops Research*, 200, 114–121. doi: 10.1016/j.fcr.2016.09.024

Presterl, T., Ouzunova, M., Schmidt, W., Möller, E. M., Röber, F. K., & Knaak, C. et al. (2007). Quantitative trait loci for early plant vigour of maize grown in chilly environments. *Theoretical and Applied Genetics*, 114, 1059–1070. doi: 10.1007/s00122-006-0499-4

Qi, F., & Zhang, F. (2020). Cell cycle regulation in the plant response to stress. *Frontiers in Plant Science*, 10, 1765. doi: 10.3389/fpls.2019.01765

Qian, Y., Ren, Q., Zhang, J., & Chen, L. (2019). Transcriptomic analysis of the maize (*Zea mays* L.) inbred line B73 response to heat stress at the seedling stage. *Gene*, 692, 68–78.

Qin, F., Kakimoto, M., Sakuma, Y., Maruyama, K., Osakabe, Y., Tran, L.-S. P., Shinozaki, K., & Yamaguchi-Shinozaki, K. (2007). Regulation and functional analysis of ZmDREB2A in response to drought and heat stresses in *Zea mays* L. *Plant Journal*, 50(1), 54–69. doi: 10.1111/j.1365-313X.2007.03034.x

Qin, F., Sakuma, Y., Li, J., Liu, Q., Li, Y., & Shinozaki, K. et al. (2004). Cloning and functional analysis of a novel DREB1/CBF transcription factor involved in cold responsive gene expression in Zea mays L. *Plant and Cell Physiology*, 45, 1042–1052. doi: 10.1093/pcp/pch118

Qin, P., Zhang, G., Hu, B., Wu, J., Chen, W., & Ren, Z. et al. (2021). Leaf-derived ABA regulates rice seed development via a transporter mediated and temperature-sensitive mechanism. *Science Advances*, 7, eabc8873. doi: 10.1126/sciadv.abc8873

Qu, C., Liu, C., Gong, X., Li, C., Hong, M., & Wang, L. et al. (2012). Impairment of maize seedling photosynthesis caused by a combination of potassium deficiency and salt stress. *Environmental and Experimental Botany*, 75, 134–141.

Quan, R., Shang, M., Zhang, H., Zhao, Y., & Zhang, J. (2004). Engineering of enhanced glycine betaine synthesis improves drought tolerance in maize. *Plant Biotechnology Journal*, 2, 477–486. doi: 10.1111/j.1467-7652.2004.00093.x

Rady, M. M. (2011). Effect of 24-epibrassinolide on growth, yield, antioxidant system and cadmium content of bean (*Phaseolus vulgaris* L.) plants under salinity and cadmium stress. *Scientia Horticulturae*, 129, 232–237.

Rahil, M., Hajjeh, H., & Qanadillo, A. (2013). Effect of saline water application through different irrigation intervals on tomato yield and soil properties. *Open Journal of Soil Science*, 3, 143–147.

Rahman, H., Pekic, S., Lazic-Jancic, V., Quarrie, S. A., Shah, S. M., & Pervez, A. et al. (2011). Molecular mapping of quantitative trait loci for drought tolerance in maize plants. *Genetics and Molecular Research*, 10, 889–901. doi: 10.4238/vol10-2gmr1139

Rahman, S., Ahmad, I., & Nafees, M. (2023). Mitigation of heavy metal stress in maize (Zea mays L.) through application of silicon nanoparticles. *Biocatalysis and Agricultural Biotechnology*, 50, 102757.

Rahman, S. U., Arif, M., Hussain, K., Arshad, M., Hussain, S., Mukhtar, T., & Razaq, A. (2015). Breeding for heat stress tolerance of maize in Pakistan. *Journal of Environmental and Agricultural Sciences*, 5, 27–33.

Rashad, R. T., & Hussien, R. A. (2014). A comparison study on the effect of some growth regulators on the nutrients content of maize plant under salinity conditions. *Annals of Agricultural Sciences*, 59, 89–94.

Rasheed, R. (2009). *Salinity and extreme temperature effects on sprouting buds of sugarcane (Saccharum officinarum l.). Some histological and biochemical studies.* Faisalabad Pakistan: University of Agriculture.

Raza, A., Mubarik, M. S., Sharif, R., Habib, M., Jabeen, W., Zhang, C., Chen, H., Chen, Z. H., Siddique, K. H. M., Zhuang, W., & Varshney, R. K. (2023). Developing drought-smart, ready-to-grow future crops. *The Plant Genome*, 16, e20279. doi: 10.1002/tpg2.20279

Raza, A., Razzaq, A., Mehmood, S. S., Zou, X., Zhang, X., & Lv, Y. et al. (2019). Impact of climate change on crops adaptation and strategies to tackle its outcome: A review. *Plants*, 8, 34. doi: 10.3390/plants8020034

Raza, A., Tabassum, J., Kudapa, H., & Varshney, R. K. (2021). Can omics deliver temperature resilient ready-to-grow crops? *Critical Reviews in Biotechnology*, 41(8), 1209–1232. doi: 10.1080/07388551.2021.1898332

Reddy, V. S., Day, I. S., Thomas, T., & Reddy, A. S. N. (2004). KIC, a novel Ca^{2+} binding protein with one EF-H and motif, interacts with a microtubule motor protein and regulates trichome morphogenesis. *Plant Cell*, 16(1), 185–200. doi: 10.1105/tpc.016600

Reimold, A. M., Etkin, A., Clauss, I., Perkins, A., Friend, D. S., Zhang, J., Horton, H. F., Scott, A., Orkin, S. H., Byrne, M. C., Grusby, M. J., & Glimcher, L. H. (2000). An essential role in liver development for transcription factor XBP-1. *Genes and Development*, 14(2), 152–157. doi: 10.1101/gad.14.2.152

Rellán-Álvarez, R., Ortega-Villasante, C., & Álvarez-Fernández, A. et al. (2006). Stress responses of *Zea mays* to cadmium and mercury. *Plant Soil*, 279, 41–50. doi: 10.1007/s11104-005-3900-1

Revilla, P., Malvar, R., Cartea, M., Butrón, A., & Ordás, A. (2000). Inheritance of cold tolerance at emergence and during early season growth in maize. *Crop Science*, 40, 1579–1585. doi: 10.2135/cropsci2000.4061579x

Rezaei, E. E., Webber, H., Gaiser, T., Naab, J., & Ewert, F. (2015). Heat stress in cereals: Mechanisms and modelling. *European Journal of Agronomy*, 64, 98–113. doi: 10.1016/j.eja.2014.10.003

Ribaut, J. M., De Vicente, M. C., & Delannay, X. (2010). Molecular breeding in developing countries: Challenges and perspectives. *Current Opinion in Plant Biology*, 13, 213–218. doi: 10.1016/j.pbi.2009.12.011

Ribaut, J. M., Jiang, C., Gonzalez-de-Leon, D., Edmeades, G. O., & Hoisington, D. A. (1997). Identification of quantitative trait loci under drought conditions in tropical maize. 1. Flowering parameters and the anthesis-silking interval. *Theoretical and Applied Genetics*, 92, 905–914. doi: 10.1007/BF00221905

Ribaut, J. M., & Ragot, M. (2007). Marker-assisted selection to improve drought adaptation in maize: The backcross approach, perspectives, limitations, and alternatives. *Journal of Experimental Botany*, 58, 351–360. doi: 10.1093/jxb/erl214

Ribeiro, C., Hennen-Bierwagen, T. A., Myers, A. M., Cline, K., Settles, A. M., & Carver, R. J. (2020). Engineering 6-phosphogluconate dehydrogenase to improve heat tolerance in maize seed development. doi: 10.1101/2020.05.21.108985

Rihan, H. Z., Al-Issawi, M., & Fuller, M. P. (2017). Advances in physiological and molecular aspects of plant cold tolerance. *Journal of Plant Interactions*, 12, 143–157. doi: 10.1080/17429145.2017.1308568

Rijavec, T., Jain, M., Dermastia, M., & Chourey, P. S. (2011). Spatial and temporal profiles of cytokinin biosynthesis and accumulation in developing caryopses of maize. *Annals of Botany*, 107, 1235–1245.

Rios-Gonzalez, K., Erdei, L., & Lips, S. H. (2002). The activity of antioxidant enzymes in maize and sunflower seedlings as affected by salinity and different nitrogen sources. *Plant Science*, 162, 923–930.

Ristic, Z., Bukovnik, U., Momčilović, I., Fu, J., & Vara Prasad, P. V. (2008). Heat-induced accumulation of chloroplast protein synthesis elongation factor, EF-Tu, in winter wheat. *Journal of Plant Physiology*, 165, 192–202. doi: 10.1016/j.jplph.2007.03.003

Ritonga, F. N., & Chen, S. (2020). Physiological and molecular mechanism involved in cold stress tolerance in plants. *Plants*, 9, 560. doi: 10.3390/plants9050560

Rivero, R. M., Mittler, R., Blumwald, E., & Zandalinas, S. I. (2022). Developing climate-resilient crops: Improving plant tolerance to stress combination. *The Plant Journal*, 109(2), 373–389. doi: 10.1111/tpj.15483

Rizvi, A., & Khan, M. S. (2018). Heavy metal induced oxidative damage and root morphology alterations of maize (Zea mays L.) plants and stress mitigation by metal tolerant nitrogen fixing Azotobacter chroococcum. *Ecotoxicology and Environmental Safety*, 157, 9–20.

Rodríguez, V., Velasco, P., Garrido, J. L., Revilla, P., Ordás, A., & Butrón, A. (2013). Genetic regulation of cold-induced albinism in the maize inbred line A661. *Journal of Experimental Botany*, 64, 3657–3667. doi: 10.1093/jxb/ert189

Romero-Puertas, M. C., Palma, J. M., Gomez, M., Del Rıo, L. A., & Sandalio, L. M. (2002). Cadmium causes the oxidative modification of proteins in pea plants. *Plant, Cell & Environment*, 25(5), 677–686.

Ruelland, E., Cantrel, C., Gawer, M., Kader, J. C., & Zachowski, A. (2002). Activation of phospholipases C and D is an early response to a cold exposure in Arabidopsis suspension cells. *Plant Physiology*, 130(2), 999–1007.

Rumbaugh, M. D., Pendery, B. M., & James, D. W. (1993). Variation in the salinity tolerance of strawberry clover (*Trifolium fragiferum* L.). *Plant Soil*, 153, 265–271.

Rupwate, S. D., & Rajasekharan, R. (2012). Plant phosphoinositide specific phospholipase C: An insight. *Plant Signaling and Behavior*, 7, 1281–1283. doi: 10.4161/psb.21436

Sabagh, A. E., Mbarki, S., Hossain, A., Iqbal, M. A., Islam, M. S., Raza, A., ... & Farooq, M. (2021). Potential role of plant growth regulators in administering crucial processes against abiotic stresses. *Frontiers in Agronomy*, 3, 648694.

Sabir, M., Waraich, E. A., Hakeem, K. R., Ozturk, M., Ahmad, H. R., & Shahid, M. (2015). Phytoremediation: Mechanisms and adaptations. In: K. Hakeem, M. Sabir, M. Ozturk, & A. Mermut, eds. *Soil remediation and plants: Prospects and challenges* (pp. 85–105). New York: Elsevier.

Sah, R. P., Chakraborty, M., Prasad, K., Pandit, M., Tudu, V. K., & Chakravarty, M. K. et al (2020). Impact of water deficit stress in maize: Phenology and yield components. *Scientific Reports*, 10, 1–5. doi: 10.1038/s41598-020-59689-7

Salama, Z. A., Shaaban, M. M., & Abou El-Nour, A. E. A. (1996). Effect of iron foliar application on increasing tolerance of maize seedlings to saline irrigation water. *Egyptian Journal of Applied Science*, 11, 169–175.

Salim, B. B. M. (2014). Effect of boron and silicon on alleviating salt stress in maize. *Middle East Journal of Agriculture Research*, 3, 1196–1204.

Salvi, S., & Tuberosa, R. (2005). To clone or not to clone plant QTLs: Present and future challenges. *Trends in Plant Science*, 10, 297–304. doi: 10.1016/j.tplants.2005.04.008

Sang, H., Guo, W., Gao, Y., Jiao, X., & Pan, X. (2020). Effects of alternating fresh and saline water irrigation on soil salinity and chlorophyll fluorescence of summer maize. *Water*, 12(11), 3054.

Schaid, D. J., Chen, W., & Larson, N. B. (2018). From genome-wide associations to candidate causal variants by statistical fine-mapping. *Nature Reviews Genetics*, 19, 491–504. doi: 10.1038/s41576-018-0016-z

Scheben, A., & Edwards, D. (2017). Genome editors take on crops. *Science*, 355, 1122–1123. doi: 10.1126/science.aal4680

Schnable, P. S., Ware, D., Fulton, R. S., Stein, J. C., Wei, F., & Pasternak, S. et al. (2009). The B73 maize genome: Complexity, diversity, and dynamics. *Science*, 326, 1112–1115. doi: 10.1126/science.1178534

Schoper, J. B., Lambert, R. J., & Vasilas, B. L. (1987a). Pollen viability, pollen shedding, and combining ability for tassel heat tolerance in maize. *Crop Science*, 27(1), 27–31.

Schubert, S., Neubert, A., Schierholt, A., Sumer, A., & Zorb, C. (2009). Development of salt resistant maize hybrids: The combination of physiological strategies using conventional breeding methods. *Plant Science*, 177, 196–202.

Schwietzke, S., Kim, Y., Ximenez, E., Mosier, N., & Ladisch, M. (2009). *Molecular genetic approaches to maize improvement*. Berlin: Springer-Verlag, 347–364.

Seetharam, K., Kuchanur, P. H., Koirala, K. B., Tripathi, M. P., Patil, A., Sudarsanam, V., Das, R. R., Chaurasia, R., Pandey, K., Vemuri, H., Vinayan, M. T., Nair, S. K., Babu, R., & Zaidi, P. H. (2021). Genomic regions associated with heat stress tolerance in tropical maize (*Zea mays* L.). *Scientific Reports*, 11(1), 137301. doi: 10.1038/s41598-021-93061-7

Setter, T. L., Yan, J., Warburton, M., Ribaut, J. M., Xu, Y., & Sawkins, M. et al. (2011). Genetic association mapping identifies single nucleotide polymorphisms in genes that affect abscisic acid levels in maize floral tissues during drought. *Journal of Experimental Botany*, 62, 701–716. doi: 10.1093/jxb/erq308

Sezer, S. (2014). Effect of boron fertilizer applications on the growth and B, N uptake of maize (*Zea mays* L.) under the different soils. *Journal of Food, Agriculture and Environment*, 12, 1323–1327.

Shahid, M., Khalid, S., Abbas, G., Shahid, N., Nadeem, M., Sabir, M., Aslam, M., & Dumat, C. (2015) Heavy metal stress and crop productivity. In: K. Hakeem et al., eds. *Crop production and global environmental issues*. Cham: Springer.

Shahzad, M., Witzel, K., Zorb, C., & Muhling, K. H. (2012). Growth-related changes in subcellular ion patterns in maize leaves (*Zea mays* L.) under salt stress. *Journal of Agronomy and Crop Science*, 198, 46–56.

Shan, Q., Wang, Y., Li, J., Zhang, Y., Chen, K., & Liang, Z. et al. (2013). Targeted genome modification of crop plants using a CRISPR-Cas system. *Nature Biotechnology*, 31, 686–688. doi: 10.1038/nbt.2650

Shao, R.-X., Yu, K.-K., Li, H.-W., Jia, S.-J., Yang, Q.-H., Zhao, X., Zhao, Y.-L., & Liu, T.-X. (2021). The effect of elevating temperature on the growth and development of reproductive organs and yield of summer maize. *Journal of Integrative Agriculture*, 20(7), 1783–1795.

Sharma, R. K., Kumar, S., Vatta, K., Bheemanahalli, R., Dhillon, J., & Reddy, K. N. (2022). Impact of recent climate change on corn, rice, and wheat in southeastern USA. *Scientific Reports*, 12(1), 16928. doi: 10.1038/s41598-022-21454-3

Shelden, M. C., & Roessner, U. (2013). Advances in functional genomics for investigating salinity stress tolerance mechanisms in cereals. *Frontiers in Plant Science*, 4, 123.

Shi, J., Gao, H., Wang, H., Lafitte, H. R., Archibald, R. L., & Yang, M. et al. (2017). ARGOS 8 variants generated by CRISPR-Cas9 improve maize grain yield under field drought stress conditions. *Plant Biotechnology Journal*, 15, 207–216. doi: 10.1111/pbi.12603

Shou, H., Bordallo, P., & Wang, K. (2004). Expression of the Nicotiana protein kinase (NPK1) enhanced drought tolerance in transgenic maize. *Journal of Experimental Botany*, 55, 1013–1019. doi: 10.1093/jxb/erh129

Silva, N. V., & Patron, N. J. (2017). CRISPR-based tools for plant genome engineering. *Emerging Topics in Life Sciences*, 1, 135–149. doi: 10.1042/ETLS20170011

Singh, S., Srivastava, P. K., Kumar, D., Tripathi, D. K., Chauhan, D. K., & Prasad, S. M. (2015). Morpho-anatomical and biochemical adapting strategies of maize (Zea mays L.) seedlings against lead and chromium stresses. *Biocatalysis and Agricultural Biotechnology*, 4(3), 286–295.

Sivakumar, S., Dhasarathan, M., Karthikeyan, A., Bharathi, P., Ganesamurthy, K., & Senthil, N. (2019). Population structure and association mapping studies for yield-related traits in maize (Zea mays L.). *Current Plant Biology*, 18, 100103.

Sowiński, P., Fronk, J., Jończyk, M., Grzybowski, M., Kowalec, P., & Sobkowiak, A. (2020). Maize response to low temperatures at the gene expression level: A critical survey of transcriptomic studies. *Frontiers in Plant Science*, 11, 576941. doi: 10.3389/fpls.2020.576941

Spanswick, R. M. (1981). Electrogenic ion pumps. *Annual Review of Plant Physiology*, 32, 267–289.

Spielbauer, G., Li, L., Römisch-Margl, L., Do, P. T., Fouquet, R., Fernie, A. R., ... & Settles, A. M. (2013). Chloroplast-localized 6-phosphogluconate dehydrogenase is critical for maize endosperm starch accumulation. *Journal of experimental botany*, 64(8), 2231–2242.

Stefanov, M. A., Rashkov, G. D., Yotsova, E. K., Borisova, P. B., Dobrikova, A. G., & Apostolova, E. L. (2021). Different sensitivity levels of the photosynthetic apparatus in *Zea mays* L. and *Sorghum bicolor* L. under salt stress. *Plants (Basel)*, 10(7), 1469. doi: 10.3390/plants10071469

Stone, P. (2001) The effects of heat stress on cereal yield and quality. In: A. S. Basara, ed. *Crop responses and adaptations to temperature stress* (pp. 243–291). Binghamton, New York, USA: Food Products Press.

Strigens, A., Freitag, N. M., Gilbert, X., Grieder, C., Riedelsheimer, C., & Schrag, T. A. et al (2013). Association mapping for chilling tolerance in elite flint and dent maize inbred lines evaluated in growth chamber and field experiments. *Plant, Cell & Environment*, 36, 1871–1887. doi: 10.1111/pce.12096

Sun, C., Ali, K., Yan, K., Fiaz, S., Dormatey, R., & Bi, Z. et al. (2021). Exploration of epigenetics for improvement of drought and other stress resistance in crops: A review. *Plants*, 10, 12–26. doi: 10.3390/plants10061226

Sun, Y., He, Y., Irfan, A. R., Liu, X., Yu, Q., & Zhang, Q. et al. (2020). Exogenous brassinolide enhances the growth and cold resistance of maize (Zea mays L.) seedlings under chilling stress. *Agronomy*, 10, 488. doi: 10.3390/agronomy10040488

Sun, Z. Q., Dong, X. Q., Wang, X. J., Zheng, D. F., Dong, L., & Liu, Z. H. (2014). Effect of saline drip irrigation to soil water and salt distribution and cotton yield in northern Shandong plain. *Agricultural Research in the Arid Areas*, 32, 12–18.

Suwa, R., Hakata, H., Hara, H., El-Shemy, H. A., Adu-Gyamfi, J. J., Nguyen, N. T., Kanai, S., Lightfoot, D. A., Mohapatra, P. K., & Fujita, K. (2010). High temperature effects on photosynthate partitioning and sugar metabolism during ear expansion in maize (*Zea mays* L.) genotypes. *Plant Physiology and Biochemistry*, 48(2–3), 124–130. doi: 10.1016/j.plaphy.2009.12.010

Suzuki, N., Bassil, E., Hamilton, J. S., Inupakutika, M. A., Zandalinas, S. I., Tripathy, D., Luo, Y., Dion, E., Fukui, G., Kumazaki, A., Nakano, R., Rivero, R. M., Verbeck, G. F., Azad, R. K., Blumwald, E., & Mittler, R. (2016). ABA is required for plant acclimation to a combination of salt and heat stress. *PLoS ONE*, 11(1), e0147625. doi: 10.1371/journal.pone.0147625

Swamy, B. P., Vikram, P., Dixit, S., Ahmed, H. U., & Kumar, A. (2011). Metaanalysis of grain yield QTL identified during agricultural drought in grasses showed consensus. *BMC Genomics*, 12, 319. doi: 10.1186/1471-2164-12-319

Swarbreck, S. M., Colaço, R., & Davies, J. M. (2013). Plant calcium permeable channels. *Plant Physiology*, 163(2), 514–522. doi: 10.1104/pp.113.220855

Taiz, L., Zeiger, E., Møller, I. M., & Murphy, A. (2015). *Plant physiology and development* (6th ed.). New York, NY: Sinauer Associates.

Tao, Z.-Q., Chen, Y.-Q., Li, C., Zou, J.-X., Yan, P., Yuan, S.-F., Wu, X., & Sui, P. (2016). The causes and impacts for heat stress in spring maize during grain filling in the North China plain – A review. *Journal of Integrative Agriculture*, 15(12), 2677–2687. https://doi.org/10.1016/S2095-3119(16)61409-0

Tas, T. (2022). Physiological and biochemical responses of hybrid maize (Zea mays L.) varieties grown under heat stress conditions. *PeerJ*, 10, e14141. doi: 10.7717/peerj.14141

Tedeschi, A., & Menenti, M. (2002). Simulation studies of long-term saline water use: Model validation and evaluation of schedules. *Agricultural Water Management*, 54, 123–157.

Tena, G., Asai, T., Chiu, W. L., & Sheen, J. (2001). Plant mitogen activated protein kinase signaling cascades. *Current Opinion in Plant Biology*, 4, 392–400.

Tesfaye, K., Kruseman, G., Cairns, J. E., Zaman-Allah, M., Wegary, D., & Zaidi, P. H. et al. (2018). Potential benefits of drought and heat tolerance for adapting maize to climate change in tropical environments. *Climate Risk Management*, 19, 106–119. doi: 10.1016/j.crm.2017.10.001

Tester, M., & Davenport, R. (2003). Na+ tolerance and Na+ transport in higher plants. *Annals of Botany*, 91, 503–507.

Thiebaut, F., Hemerly, A. S., & Ferreira, P. C. G. (2019). A role for epigenetic regulation in the adaptation and stress responses of non-model plants. *Frontiers in Plant Science*, 10, 246. doi: 10.3389/fpls.2019.00246

Thind, S., Chaudhary, M. S., Ditta, A., Hussain, I., Parveen, A., & Ullah, N. et al. (2022). Impact of mycorrhizal fungi from different rhizospheric soils on fungal colonization, growth, and chlorophyll contents of Cenchrus ciliaris. *Agronomy*, 12, 2644.

Thirunavukkarasu, N., Hossain, F., Arora, K., Sharma, R., Shiriga, K., & Mittal, S. et al. (2014). Functional mechanisms of drought tolerance in subtropical maize (Zea mays L.) identified using genome-wide association mapping. *BMC Genomics*, 15, 1182. doi: 10.1186/1471-2164-15-1182

Thomson, L. M. (1966). Weather variability, climate change and grain production. *Science*, 188(4188), 535–541.

Thornsberry, J. M., Goodman, M. M., Doebley, J., Kresovich, S., Nielsen, D., & Buckler, E. S. (2001). Dwarf8 polymorphisms associate with variation in flowering time. *Nature Genetics*, 28, 286–289. doi: 10.1038/90135

Tondelli, A., Francia, E., Barabaschi, D., Aprile, A., Skinner, J. S., & Stockinger, E. J. et al. (2006). Mapping regulatory genes as candidates for cold and drought stress tolerance in barley. *Theoretical and Applied Genetics*, 112, 445–454. doi: 10.1007/s00122-005-0144-7

Trachsel, S., Sun, D., SanVicente, F. M., Zheng, H., Atlin, G. N., & Suarez, E. A. et al. (2016). Identification of QTL for early vigor and stay-green conferring tolerance to drought in two connected advanced backcross populations in tropical maize (Zea mays L.). *PLoS One*, 11, e0149636. doi: 10.1371/journal.pone.0149636

Tuberosa, R., & Salvi, S. (2006). Genomics-based approaches to improve drought tolerance of crops. *Trends in Plant Science*, 11, 405–412. doi: 10.1016/j.tplants.2006.06.003.

Tuberosa, R., Salvi, S., Sanguineti, M. C., Landi, P., Maccaferri, M., & Conti, S. (2002). Mapping QTLs regulating morpho-physiological traits and yield: Case studies, shortcomings and perspectives in drought-stressed maize. *Annals of Botany*, 89, 941–963. doi: 10.1093/aob/mcf134

Tuna, A. L., Kaya, C., Dikilitas, M., & Higgs, D. (2008). The combined effects of gibberellic acid and salinity on some antioxidant enzyme activities, plant growth parameters and nutritional status in maize plants. *Environmental and Experimental Botany*, 62, 1–9.

Tuna, A. L., Kaya, C., Dikilttas, M., Yokas, I., Burun, B., & Altunlu, H. (2007). Comparative effects of various salicylic acid derivatives on key growth parameters and some enzyme activities in salinity stressed maize (*Zea mays* L.) plants. *Pakistan Journal of Botany*, 39, 787–798.

Turan, M. A., Elkarim, A. H., Taban, N., & Taban, S. (2009). Effect of salt stress on growth, stomatal resistance, proline and chlorophyll concentrations on maize plant. *African Journal of Agricultural Research*, 4, 893–897.

Turk, H., Erdal, S., & Dumlupinar, R. (2019). Carnitine-induced physiobiochemical and molecular alterations in maize seedlings in response to cold stress. *Archives of Agronomy and Soil Science*, 66, 925–941. doi: 10.1080/03650340.2019.1647336

Urban, J., Ingwers, M., McGuire, M. A., & Teskey, R. O. (2017). Stomatal conductance increases with rising temperature. *Plant Signaling & Behavior*, 12(8), e1356534.

Valliyodan, B., & Nguyen, H. T. (2006). Understanding regulatory networks and engineering for enhanced drought tolerance in plants. *Current Opinion in Plant Biology*, 9, 189–195. doi: 10.1016/j.pbi.2006.01.019

Vargas, M., van Eeuwijk, F. A., Crossa, J., & Ribaut, J. M. (2006). Mapping QTLs and QTL x environment interaction for CIMMYT maize drought stress program using factorial regression and partial least squares methods. *Theoretical and Applied Genetics*, 112, 1009–1023. doi: 10.1007/s00122-005-0204-z

Vaughan, M. M., Block, A., Christensen, S. A., Allen, L. H., & Schmelz, E. A. (2018). The effects of climate change associated abiotic stresses on maize phytochemical defenses. *Phytochemistry Reviews*, 17, 37–49.

Verkest, A., Byzova, M., Martens, C., Willems, P., Verwulgen, T., & Slabbinck, B. et al. (2015). Selection for improved energy use efficiency and drought tolerance in canola results in distinct transcriptome and epigenome changes. *Plant Physiology*, 168, 1338–1350. doi: 10.1104/pp.15.00155

Verslues, P. E., Agarwal, M., Katiyar-Agarwal, S., Zhu, J., & Zhu, J.-K. (2006). Methods and concepts in quantifying resistance to drought, salt and freezing, abiotic stresses that affect plant water status. *The Plant Journal*, 45, 523–539. doi: 10.1111/j.1365-313X.2005.02593.x

Vinocur, B., & Altman, A. (2005). Recent advances in engineering plant tolerance to abiotic stress: Achievements and limitations. *Current Opinion in Biotechnology*, 16, 123–132. doi: 10.1016/j.copbio.2005.02.001

Vitale, A., & Boston, R. S. (2008). Endoplasmic reticulum quality control and the unfolded protein response: Insights from plants. *Traffic*, 99, 1581–1588. doi: 10.1111/j.1600-0854.2008.00780.x

Waititu, J. K., Cai, Q., Sun, Y., Sun, Y., Li, C., & Zhang, C. et al. (2021). Transcriptome profiling of maize (Zea mays L.) leaves reveals key cold-responsive genes, transcription factors, and metabolic pathways regulating cold stress tolerance at the seedling stage. *Genes*, 12, 1638.

Walker, J. M. (1969). One-degree increments in soil temperatures affect maize seedling behavior. *Soil Science Society of America Journal*, 33(5), 729–736.

Wallace, J. G., Zhang, X., Beyene, Y., Semagn, K., Olsen, M., & Prasanna, B. M. et al (2016). Genome-wide association for plant height and flowering time across 15 tropical maize populations under managed drought stress and well-watered conditions in sub-Saharan Africa. *Crop Science*, 56, 2365–2378. doi: 10.2135/cropsci2015.10.0632

Wang, B., Zhang, J., Xia, X., & Zhang, W. H. (2011). Ameliorative effect of brassinosteroid and ethylene on germination of cucumber seeds in the presence of sodium chloride. *Plant Growth Regulators*, 65, 407–413.

Wang, C., & Dong, Y. (2009). Overexpression of maize ZmDBP3 enhances tolerance to drought and cold stress in transgenic Arabidopsis plants. *Biologia*, 64, 1108–1114. doi: 10.2478/s11756-009-0198-0

Wang, C. R., Yang, A. F., Yue, G. D., Gao, Q., Yin, H. Y., & Zhang, J. R. (2008). Enhanced expression of phospholipase C 1 (ZmPLC1) improves drought tolerance in transgenic maize. *Planta*, 227, 1127–1140. doi: 10.1007/s00425-007-0686-9

Wang, Q., Xu, J., Pu, X., Lv, H., Liu, Y., & Ma, H. et al. (2021). Maize DNA methylation in response to drought stress is involved in target gene expression and alternative splicing. *International Journal of Molecular Science*, 22, 82–85. doi: 10.3390/ijms22158285

Wang, Q. M., Huo, Z. L., Zhang, L. D., Wang, J. H., & Zhao, Y. (2016). Impact of saline water irrigation on water use efficiency and soil salt accumulation for spring maize in arid regions of China. *Agricultural Water Management*, 163, 125–138.

Wang, T., Xu, Z., & Pang, G. (2019). Effects of irrigating with brackish water on soil moisture, soil salinity, and the agronomic response of winter wheat in the Yellow River Delta. *Sustainability*, 11(20), 5801.

Wang, T., Zhang, H., & Zhu, H. (2019). CRISPR technology is revolutionizing the improvement of tomato and other fruit crops. *Horticulture Research*, 6, 1–13. doi: 10.1038/s41438-019-0159-x

Wang, X., Feng, J., White, P. J., Shen, J., & Cheng, L. (2020a). Heterogeneous phosphate supply influences maize lateral root proliferation by regulating auxin redistribution. *Annals of Botany*, 125, 119–130. doi: 10.1093/aob/mcz154

Wang, X., Yan, Y., Xu, C., Wang, X., Luo, N., Wei, D., Meng, Q., & Wang, P. (2021a). Mitigating heat impacts in maize (*Zea mays* L.) during the reproductive stage through biochar soil amendment. *Agriculture, Ecosystems & Environment*, 311, 107321.

Wang, Z., Jia, Q., Zhang, B., Ming, B., Bai, L., Wang, F., Wang, Y., Yu, S., Zou, R., & Wang Z. (2026). Unveiling the physiological basis of cold tolerance in maize: Root architecture, photosynthetic stability, and pod-mediated defense under delayed chilling stress. *Plants (Basel)*, 15(3), 517. doi: 10.3390/plants15030517

Wang, X., Zenda, T., Liu, S., Liu, G., Jin, H., & Dai, L. et al. (2019). Comparative proteomics and physiological analyses reveal important maize filling-kernel drought-responsive genes and metabolic pathways. *International Journal of Molecular Sciences*, 20, 3743. doi: 10.3390/ijms20153743

Wang, Y., Chu, J., Sun, Z., & Wang, D. (2006). Study on physiological response of maize to low temperature and comparison of frost hardiness varieties. *Chinese Agricultural Science Bulletin*, 9, 210–212. doi: 10.3969/j.issn.1000-6850.2006.09.052

Wang, Y., Wen, T., Hu, J., Han, R., Zhu, Y., & Guan, Y. et al. (2013). Relationship between endogenous salicylic acid and antioxidant enzyme activities in maize seedlings under chilling stress. *Experimenta. Agriculture*, 49, 295–308.

Wang, Z., Li, G., Sun, H., Ma, L., Guo, Y., & Zhao, Z. et al. (2018). Effects of drought stress on photosynthesis and photosynthetic electron transport chain in young apple tree leaves. *Biology Open*, 7, bio035279. doi: 10.1242/bio.035279

Wang, H., An, T., Huang, D., Liu, R., Xu, B., Zhang, S., ... & Chen, Y. (2021). Arbuscular mycorrhizal symbioses alleviating salt stress in maize is associated with a decline in root-to-leaf gradient of Na+/K+ ratio. *BMC Plant Biology*, 21(1), 457.

Waraich, E. A., Ahmad, R., Halim, A., & Aziz, T. (2012). Alleviation of temperature stress by nutrient management in crop plants: A review. *Journal of Soil Science and Plant Nutrition*, 12, 221–244. doi: 10.4067/S0718-95162012000200003

Weckwerth, P., Ehlert, B., & Romeis, T. (2015). ZmCPK1, a calcium independent kinase member of the Zea mays CDPK gene family, functions as a negative regulator in cold stress signalling. *Plant, Cell & Environment*, 38, 544–558. doi: 10.1111/pce.12414

Wei, S., Luo, N., Jing, L. I., Wanrong, G. U., Chen, G., & Lijie, L. I. et al. (2014). Change of the root protective enzyme activities and endogenous hormones of maize seedling under low-temperature stress. *Journal of Northeast Agricultural University*, 45, 1–8.

Wojtyla, Ł., Paluch-Lubawa, E., Sobieszczuk-Nowicka, E., & Garnczarska, M. (2020). Drought stress memory and subsequent drought stress tolerance in plants. In: *Priming-mediated stress and cross-stress tolerance in crop plants* (pp. 115–131). London: Academic Press.

Wu, J., Jiang, Y., Liang, Y., Chen, L., Chen, W., & Cheng, B. (2019). Expression of the maize MYB transcription factor ZmMYB3R enhances drought and salt stress tolerance in transgenic plants. *Plant Physiology and Biochemistry*, 137, 179–188. doi: 10.1016/j.plaphy.2019.02.010

Wu, S., Ning, F., Zhang, Q., Wu, X., & Wang, W. (2017). Enhancing omics research of crop responses to drought under field conditions. *Frontiers in Plant Science*, 8, 174. doi: 10.3389/fpls.2017.00174

Würschum, T., Weiß, T. M., Renner, J., Utz, H. U., Gierl, A., Jonczyk, R., Römisch-Margl, L., Schipprack, W., Schön, C.-C., Schrag, T. A., Leiser, W. L., & Melchinger, A. E. (2022). High-resolution association mapping with libraries of immortalized lines from ancestral landraces. *Theoretical and Applied Genetics*, 135(1), 243–256. doi: 10.1007/s00122-021-03963-3

Xia, Z., Liu, Q., Wu, J., & Ding, J. (2012). ZmRFP1, the putative ortholog of SDIR1, encodes a RING-H2 E3 ubiquitin ligase and responds to drought stress in an ABA-dependent manner in maize. *Gene*, 495, 146–153. doi: 10.1016/j.gene.2011.12.028

Xiang, B., Li, C., Zhang, J., Luo, Y., & Jiang, A. (2012). Transformation of CBF1 gene to maize grass SAUMZ1 improves its ability of cold resistance. *Pratacultural Science*, 29, 1374–1378.

Xu, H., Twine, T. E., & Girvetz, E. (2016). Climate change and maize yield in Iowa. *PLoS ONE*, 11, e0156083.

Yadava, P., Abhishek, A., Singh, R., Singh, I., Kaul, T., & Pattanayak, A. et al. (2017). Advances in maize transformation technologies and development of transgenic maize. *Frontier in Plant Science*, 7, 1949. doi: 10.3389/fpls.2016.01949

Yan, J., Wu, Y., Li, W., Qin, X., Wang, Y., & Yue, B. (2017). Genetic mapping with test-crossing associations and F2: 3 populations reveals the importance of heterosis in chilling tolerance at maize seedling stage. *Scientific Reports*, 7, 3232. doi: 10.1038/s41598-017-03585-0

Yan, Q., Zong, X., Wu, F., Li, J., Ma, T., Zhao, Y., & Zhang, J. (2020). Integrated analysis of co-expression, conserved genes and gene families reveal core regulatory network of heat stress response in Cleistogenes songorica, a xerophyte perennial desert plant. *BMC Genomics*, 21(1), 715.

Yang, X., & Lu, C. (2005). Photosynthesis is improved by exogenous glycinebetaine in salt-stressed maize plants. *Physiologia Plantarum*, 124(3), 343–352.

Yang, D., Sun, Y., Irfan, A. R., Liu, X., Jinying, L. V., & Juping, Y. U. et al. (2018). Effect of low temperature stress on germination and physiological of maize seedling. *Journal of Northeast Agricultural University*, 49, 8–44. do

Yang, G., Zou, H., Wu, Y., Liu, H., & Yuan, Y. (2011). Identification and characterization of candidate genes involved in chilling responses in maize (Zea mays L.). *Plant Cell, Tissue and Organ Culture*, 106, 127–141. doi: 10.1007/s11240-010-9900-8

Yang, H., Gu, X., Ding, M., Lu, W., & Lu, D. (2018). Heat stress during grain filling affects activities of enzymes involved in grain protein and starch synthesis in waxy maize. *Scientific Reports*, *8*(1), 15665. doi: 10.1038/s41598-018-33644-z

Yang, H., Huang, T., Ding, M., Lu, D., & Lu, W. (2017). High temperature during grain filling impacts on leaf senescence in waxy maize. *Agronomy Journal*, 109(3), 906–916. doi: 10.2134/agronj2016.08.0452

Yang, J., Mao, J., Yu, Y., Li, C., Wang, Y., & Hu, J. (2016). Effects of chilling on antioxidant enzyme activity and related gene expression levels during seed germination. *Journal of Nuclear Agricultural Sciences*, 30, 1840–1847.

Yi, Q., Alvarez-Iglesias, L., Malvar, R. A., Romay, M. C., & Revilla, P. (2021). A worldwide maize panel revealed new genetic variation for cold tolerance. *Theoretical and Applied Genetics*, 134, 1083–1094. doi: 10.1007/s00122-020-03753-3

Yilmaz, R., Sakcali, S., Yarci, C., Aksoy, A., & Ozturk, M. (2006). Use of *Aesculus hippocastanum* L. as a biomonitor of heavy metal pollution. *Pakistan Journal of Botany*, 38(5), 1519–1527.

Yu, D., Ke, L., Zhang, D., Wu, Y., Sun, Y., & Mei, J. et al. (2021). Multi-omics assisted identification of the key and species-specific regulatory components of drought-tolerant mechanisms in Gossypium stocksii. *Plant Biotechnology Journal*, 19, 1690. doi: 10.1111/pbi.13655

Yu, H., Qu, J., Guo, X., Li, L., Zhang, X., Yang, Q., ... & Fu, F. (2022). Overexpression of vacuolar H^+-pyrophosphatase (H^+-PPase) gene from Ammopiptanthus nanus enhances drought tolerance in maize. *Journal of Agronomy and Crop Science*, 208(5), 633–644.

Yuan, Y., Cairns, J. E., Babu, R., Gowda, M., Makumbi, D., Magorokosho, C., ... & Zhang, X. (2019). Genome-wide association mapping and genomic prediction analyses reveal the genetic architecture of grain yield and flowering time under drought and heat stress conditions in maize. *Frontiers in Plant Science*, 9, 1919.

Yue, R., Lu, C., Sun, T., Peng, T., Han, X., Qi, J., Yan, S., & Tie, S. (2015). Identification and expression profiling analysis of calmodulin-binding transcription activator genes in maize (*Zea mays* L.) under abiotic and biotic stresses. *Frontiers in Plant Science*, 6, 576. doi: 10.3389/fpls.2015.00576

yyGreaves, J. A. (1996). Improving suboptimal temperature tolerance in maize-the search for variation. *Journal of Experimental Botany*, 47(3), 307–323.

Zaidi, P., Yadav, M., Maniselvan, P., Khan, R., Shadakshari, T., Singh, R., & Pal, D. (2010). Morpho physiological traits associated with cold stress tolerance in tropical maize (*Zea mays* L.). *Maydica*, 55, 201–208.

Zaidi, P. H., & Singh, N. N. (2010). Effect of waterlogging on growth, biochemical compositions and reproduction in maize. *Journal of Plant Biology*, 28, 61–69.

Zelman, A. K., Dawe, A., Gehring, C., & Berkowitz, G. A. (2012). Evolutionary and structural perspectives of plant cyclic nucleotide-gated cation channels. *Frontiers in Plant Science*, 3, 95. doi: 10.3389/fpls.2012.00095

Zeng, R., Li, Z., Shi, Y., Fu, D., Yin, P., & Cheng, J. et al. (2021). Natural variation in a type-A response regulator confers maize chilling tolerance. *Nature Communications*, 12, 4713. doi: 10.1038/s41467-021-25001-y

Zhai, S., Gao, Q., Liu, X., Sui, Z., & Zhang, J. (2013). Overexpression of a Zea mays phospholipase C1 gene enhances drought tolerance in tobacco in part by maintaining stability in the membrane lipid composition. *Plant Cell, Tissue and Organ Culture*, 115(2), 253–262.

Zhang, F., Wu, J., Sade, N., Wu, S., Egbaria, A., & Fernie, A. R. et al. (2021). Genomic basis underlying the metabolome-mediated drought adaptation of maize. *Genome Biology*, 22, 1–26. doi: 10.1186/s13059-021-02481-1

Zhang, H., Lang, Z., & Zhu, J.-K. (2018). Dynamics and function of DNA methylation in plants. *Nature Reviews Molecular Cell Biology*, 19, 489–506. doi: 10.1038/s41580-018-0016-z

Zhang, H., Li, G., Fu, C., Duan, S., Hu, D., & Guo, X. (2020a). Genome wide identification, transcriptome analysis and alternative splicing events of Hsf family genes in maize. *Scientific Reports*, 10(1), 8073. doi: 10.1038/s41598-020-65068-z

Zhang, H., Li, G., Hu, D., Zhang, Y., Zhang, Y., Shao, H., Zhao, L., Yang, R., & Guo, X. (2020b). Functional characterization of maize heat shock transcription factor gene ZmHsf01 in thermotolerance. *PeerJ*, 8, e8926. doi: 10.7717/peerj.8926

Zhang, H., Zhang, J., Xu, Q., Wang, D., Di, H., & Huang, J. et al. (2020). Identification of candidate tolerance genes to low-temperature during maize germination by GWAS and RNA-seq approaches. *BMC Plant Biology*, 20, 333. doi: 10.1186/s12870-020-02543-9

Zhang, L., Yang, P., Zheng, W., Wang, C., Zhang, C., & Niu, M. (2018). Effects of drip irrigation models on chemical clogging under saline water use in Hetao District, China. *Water*, 10, 345.

Zhang, S., Li, N., Gao, F., Yang, A., & Zhang, J. (2010). Over-expression of TsCBF1 gene confers improved drought tolerance in transgenic maize. *Molecular Breeding*, 26, 455–465. doi: 10.1007/s11032-009-9385-5

Zhang, S.-S., Yang, H., Ding, L., Song, Z.-T., Ma, H., Chang, F., & Liu, J.-X. (2017). Tissue-specific transcriptomics reveals an important role of the unfolded protein response in maintaining fertility upon heat stress in *Arabidopsis*. *Plant Cell*, 29(5), 1007–1023. doi: 10.1105/tpc.16.00916

Zhang, X. (2011). Effect of endogenous hormones content changing on the chilling tolerance in maize seed germination. *Journal of Shenyang Agricultural University*, 42, 147–151. doi: 10.3969/j.issn.1000-1700.2011.02.004

Zhang, X. (2014). Effect of low temperature on cold resistance of maize in seed germination. *Horticultural Seeds*, 9, 23–29. doi: 10.3969/j.issn.2095-0896.2014.03.009

Zhang, J., Jia, W., Yang, J., & Ismail, A. M. (2006). Role of ABA in integrating plant responses to drought and salt stresses. *Field Crops Research*, 97(1), 111–119.

Zhang, X., Mi, Y., Mao, H., Liu, S., Chen, L., & Qin, F. (2020). Genetic variation in ZmTIP1 contributes to root hair elongation and drought tolerance in maize. *Plant Biotechnology Journal*, 18, 1271–1283. doi: 10.1111/pbi.13290

Zhang, X., Zhang, B., Li, M. J., Yin, X. M., Huang, L. F., Cui, Y. C., ... & Xia, X. (2016). OsMSR15 encoding a rice C2H2-type zinc finger protein confers enhanced drought tolerance in transgenic Arabidopsis. *Journal of Plant Biology*, 59(3), 271–281.

Zhang, Y., Liang, Z., Zong, Y., Wang, Y., Liu, J., & Chen, K. et al. (2016). Efficient and transgene-free genome editing in wheat through transient expression of CRISPR/Cas9DNA or RNA. *Nature Communications*, 7, 1–8. doi: 10.1038/ncomms12617

Zhang, Y., Liu, P., Wang, C., Zhang, N., Zhu, Y., & Zou, C. et al. (2021). Genome-wide association study uncovers new genetic loci and candidate genes underlying seed chilling-germination in maize. *PeerJ*. 9, e11707. doi: 10.7717/peerj.11707

Zhang, Z., Wei, J., Li, J., Jia, Y., Wang, W., Li, J., & Gao, M. (2022). The impact of climate change on maize production: Empirical findings and implications for sustainable agricultural development. *Frontiers in Environmental Science*, 10, 954940.

Zhao, C., Liu, B., Piao, S., Wang, X., Lobell, D. B., Huang, Y., Huang, M., Yao, Y., Bassu, S., Ciais, P., Durand, J. L., Elliott, J., Ewert, F., Janssens, I. A., Li, T., Lin, E., Liu, Q., Martre, P., Müller, C., & Asseng, S. (2017). Temperature increase reduces global yields of major crops in four independent estimates. *Proceedings of the National Academy of Sciences of the United States of America*, 114(35), 9326–9331. doi: 10.1073/PNAS.1701762114

Zhao, C. X., Guo, L. Y., Jaleel, C. A., Shao, H. B., & Yang, H. B. (2008). Prospective for applying molecular and genetic methodology to improve wheat cultivars in drought environments. *Comptes Rendus Biologies*, 331, 579–586. doi: 10.1016/j.crvi.2008.05.006

Zhao, X., Ge, S., Wei, Y., Xu, X., Ding, D., & Liu, M. et al. (2020). Analysis of root physiology and related gene expression in maize (Zea mays) under low temperature stress. *Journal of Agricultural Biotechnology*, 28, 32–41.

Zhao, X., Peng, Y., Zhang, J., Fang, P., & Wu, B. (2018). Identification of QTLs and meta-QTLs for seven agronomic traits in multiple maize populations under well-watered and water-stressed conditions. *Crop Science*, 58, 507–520. doi: 10.2135/cropsci2016.12.0991

Zhao, Y., Du, H., Wang, Y., Wang, H., Yang, S., Li, C., Chen, N., Yang, H., Zhang, Y., Zhu, Y., Yang, L., & Hu, X. (2021). The calcium-dependent protein kinase ZmCDPK7 functions in heat-stress tolerance in maize. *Journal of Integrative Plant Biology*, 63(3).

Zhao, Z., Zheng, H., Wang, M., Guo, Y., Wang, Y., & Zheng, C. et al. (2023). Reshifting Na+ from shoots into long roots is associated with salt tolerance in two contrasting inbred maize (*Zea mays* L.) lines. *Plants*, 12, 1952. doi: 10.3390/plants12101952

Zhou, P., Enders, T. A., Myers, Z. A., Magnusson, E., Crisp, P. A., & Noshay, J. M. et al. (2022). Prediction of conserved and variable heat and cold stress response in maize using cis-regulatory information. *Plant Cell*, 34, 514–534. doi: 10.1093/plcell/koab267

Zhu, C., Naqvi, S., Gomez-Galera, S., Pelacho, A. M., Capell, T., & Christou, P. (2007). Transgenic strategies for the nutritional enhancement of plants. *Trends in Plant Science*, 12, 548–555.

Zhu, J., Song, N., Sun, S., Yang, W., Zhao, H., Song, W., & Lai, J. (2016). Efficiency and inheritance of targeted mutagenesis in maize using CRISPR-Cas9. *Journal of Genetics and Genomics*, *43*(1), 25–26. doi: 10.1016/j.jgg.2015.10.006

Zhu, J. J., Wang, X. P., Sun, C. X., Zhu, X. M., Meng, L. I., & Zhang, G. D. et al. (2011). Mapping of QTL associated with drought tolerance in a semi-automobile rain shelter in maize (Zea mays L.). *Agricultural Sciences in China*, 10, 987–996.

Zhu, Z., Zhang, F., Hu, H., Bakshi, A., Robinson, M. R., & Powell, J. E. et al. (2016). Integration of summary data from GWAS and eQTL studies predicts complex trait gene targets. *Nature Genetics*, 48, 481–487. doi: 10.1038/ng.3538

Zinn, K. E., Tunc-Ozdemir, M., & Harper, J. F. (2010). Temperature stress and plant sexual reproduction: Uncovering the weakest links. *Journal of Experimental Botany*, 61(7), 1959–1968. doi: 10.1093/jxb/erq053

Zizinga, A., Mwanjalolo, J. G. M., Tietjen, B., Bedadi, B., Pathak, H., Gabiri, G., & Beesigamukama, D. (2022). Climate change and maize productivity in Uganda: Simulating the impacts and alleviation with climate smart agriculture practices. *Agricultural Systems*, 199, 103407.

Zörb, C., Geilfus, C. M., Mühling, K. H., & Ludwig-Müller, J. (2013). The influence of salt stress on ABA and auxin concentrations in two maize cultivars differing in salt resistance. *Journal of Plant Physiology*, 170, 220–224.

5 What Is Low Temperature for Maize?

5.1 INTRODUCTION

Maize (*Zea mays* L.) is one of the world's most widely cultivated cereal crops, playing a vital role in global food, feed, and industrial systems. Originating in the warm tropics of Mesoamerica, maize is physiologically suited to warm growing conditions, with optimal growth typically occurring between 25°C and 30°C (Hatfield & Prueger, 2015). Consequently, maize is inherently sensitive to low-temperature stress, which can significantly impair growth, development, and yield, especially when exposure occurs at sensitive developmental stages. Low-temperature stress refers to temperatures below the optimal range for maize development and includes two main types: chilling stress (0–15°C), which causes sub-lethal but physiologically disruptive effects, and freezing stress (<0°C), which can cause cellular injury due to ice crystal formation and membrane rupture (Greaves, 1996; Prasad et al., 2017). Even in the absence of freezing, chilling temperatures can inhibit germination, slow seedling emergence, reduce photosynthetic efficiency, and alter hormonal balance (Huner et al., 1998; Fracheboud et al., 1999). The expansion of maize cultivation into temperate regions and high-altitude zones has brought new challenges, as plants often face cold spells during early spring or late autumn. Such stress events can negatively impact stand establishment, vegetative vigor, flowering synchrony, and grain fill (Tollenaar & Lee, 2002; Liu et al., 2013). As climate variability increases the likelihood of extreme temperature fluctuations, understanding how maize responds to suboptimal temperatures is becoming even more critical. This chapter aims to define low-temperature stress in the context of maize, outline its physiological and developmental impacts, and highlight genotypic variability and breeding efforts for cold tolerance. A comprehensive understanding of these aspects is essential for developing climate-resilient maize cultivars and optimizing management practices in cooler agroecosystems.

5.2 TEMPERATURE AND MAIZE

Maize (*Zea mays* L.), originally domesticated in tropical and subtropical regions, possesses a relatively high thermal threshold for germination (Greaves, 1996). This evolutionary origin contributes to its intrinsic sensitivity to low temperatures, particularly during early developmental stages such as germination (Rodríguez et al., 2010). As a result, maize cultivation is typically limited in high-latitude and high-altitude regions, where colder conditions prevail. When grown under such suboptimal thermal environments, maize exhibits slower growth rates and a shortened growing season, often resulting in reduced seedling vigor and lower yields (Hope et al., 1992;

DOI: 10.1201/9781003634027-5

Verheul et al., 1996). The critical minimum temperature for maize seed germination is approximately 10°C. At temperatures ranging between 6°C and 8°C, physiological and cellular damage becomes irreversible, significantly inhibiting germination and halting seedling development (Greaves, 1996). Consequently, 10°C is frequently used as the benchmark for screening maize germplasm for cold tolerance (Li et al., 2018). In recent years, increasing climate variability has led to more frequent and unpredictable episodes of low temperatures, particularly during the germination phase and early seedling growth. Such cold stress not only hampers germination rates and seedling vigor but also predisposes young plants to soil-borne pathogenic infections, thereby exacerbating yield losses. This underscores the pressing need to identify and characterize genetic determinants conferring cold-tolerant germination in maize. Germination represents the most temperature-sensitive phase of the maize life cycle. The optimal thermal range for maize seed germination lies between 25°C and 30°C, within which enzymatic and metabolic processes proceed efficiently. Below 10°C, seed metabolism declines markedly, leading to delayed or failed germination (Rasmussen et al., 1998). Conversely, excessive heat (above 35°C) can also impair key physiological processes, including enzymatic activities involved in water uptake and radicle emergence (Bewley et al., 2013).

Following germination, seedling emergence and establishment are also temperature-dependent, as thermal conditions regulate cell expansion, root elongation, and initial leaf development. Optimal daytime temperatures in the range of 20–30°C promote robust seedling growth, whereas exposure to lower temperatures can suppress development and increase susceptibility to pathogens (Hatfield & Prueger, 2015). Cold stress can compromise membrane fluidity, restricting nutrient uptake and impeding overall growth (Foyer et al., 2002). During the vegetative growth phase, temperature continues to influence key physiological processes, including the rate of leaf emergence, chlorophyll biosynthesis, photosynthetic efficiency, and dry matter accumulation. Moderate temperatures (25–30°C) are optimal for chlorophyll production and net photosynthetic activity (Crafts-Brandner & Salvucci, 2002). However, elevated temperatures (>35°C) can negatively affect stomatal conductance and disrupt carbon assimilation pathways, while low temperatures decelerate metabolic processes, reducing biomass production (Leipner et al., 2008). The reproductive phase—including tasseling, silking, and pollination—is especially vulnerable to temperature extremes. Fluctuations during this stage can critically affect pollen viability, silk receptivity, and fertilization success, further threatening yield stability in maize grown under temperature-stressed conditions. Ideal temperatures for maize growth and development typically range between 21°C and 32°C. Deviations from this range, particularly during the reproductive stage, can severely compromise crop performance.

Exposure to high temperatures—especially those exceeding 35°C—can drastically reduce pollen viability, impair silk receptivity, and induce pollen-silk asynchrony. These effects collectively lead to poor fertilization and increased kernel abortion (Herrero & Johnson, 1981; Schoper et al., 1986). However, suboptimal cool temperatures can delay floral development, thereby reducing the likelihood of successful fertilization.

Temperature also plays a crucial role in determining the duration and effectiveness of the grain filling period (Table 5.1). Moderate temperatures (20–30°C) during this phase

TABLE 5.1
Temperature Thresholds for Key Developmental Stages in Maize

Growth Stage	Optimal Temp (°C)	Low Temp Threshold (°C)	Effect of Low Temperature
Germination	25–30	<10	Poor/failed germination, delayed emergence
Seedling Establishment	20–28	<12	Reduced growth, leaf yellowing, increased mortality
Vegetative Growth	25–30	<15	Reduced photosynthesis, slow leaf expansion
Reproductive Stage	21–32	<18	Delayed flowering, pollen sterility, low kernel set
Grain Filling	20–30	<16	Prolonged filling, poor starch accumulation

favor assimilate accumulation and promote optimal kernel development. However, elevated temperatures can hasten plant senescence, curtail the grain filling period, and reduce final kernel size due to diminished carbon assimilation and storage capacity (Badu-Apraku et al., 1983). Conversely, low temperatures, while potentially extending the grain filling duration, often impair starch biosynthesis and increase the likelihood of fungal contamination, further compromising grain quality (Dupuis & Dumas, 1990). The cumulative impact of temperature stress throughout the maize life cycle—from germination to grain maturation—can substantially diminish both yield and grain quality. Heat stress during vegetative and reproductive stages reduces photosynthetic efficiency, disrupts hormonal signaling pathways, and increases oxidative damage at the cellular level (Wahid et al., 2007). Cold temperatures similarly inhibit physiological processes by slowing leaf expansion, delaying phenological transitions, and leading to incomplete grain filling and suboptimal ear development (Greaves, 1996).

To mitigate these adverse effects, breeding strategies for temperature resilience in maize have increasingly focused on the incorporation of adaptive traits such as early vigor, cold tolerance during seedling establishment, and heat tolerance during reproductive development (Reynolds et al., 2010). Recent advances in molecular and physiological screening technologies have facilitated the identification and selection of genotypes with enhanced resilience to both thermal extremes, thereby supporting the development of climate-resilient maize cultivars.

5.3 TEMPERATURE REQUIREMENTS

The rate of plant growth and development is fundamentally regulated by ambient temperature, with each species having a defined thermal range characterized by minimum, optimum, and maximum thresholds. These temperature parameters vary by crop and have been extensively compiled for major grain and fruit species by Hatfield et al. (2008, 2011). Projections by the Intergovernmental Panel on Climate Change (IPCC, 2007) indicate that global temperatures are expected to rise by 2–3°C over the next 30–50 years. In addition to gradual warming, the frequency, duration, and intensity of extreme temperature events—such as heatwaves—are anticipated to increase significantly (Meehl et al., 2007). Such extremes may manifest as short-term spikes of more than 5°C above average temperatures, potentially lasting several days. These events, particularly when occurring during the critical summer growth stages, could severely affect plant productivity. Despite the seriousness of this threat, relatively little research has focused specifically on the biological consequences of extreme temperature fluctuations in crops. Kumudini et al. (2014) emphasized the limited documentation on the physiological and agronomic impacts of short-term temperature extremes, which are likely to become more common. A comprehensive review by Barlow et al. (2015) on wheat (*Triticum aestivum* L.) demonstrated that both frost and excessive heat adversely affect reproductive success—frost leading to floral sterility and grain abortion, while heat stress reduced grain number and shortened the grain filling period. Further analysis by Meehl et al. (2007) revealed that daily minimum temperatures are projected to rise more rapidly than maximum temperatures. This asymmetric warming will result in increased daily mean temperatures and an elevated probability of extreme weather events. These thermal changes, particularly during sensitive growth stages, could have substantial negative effects on grain yield and overall crop productivity.

Given the magnitude and inevitability of these climatic shifts, it is imperative to deepen our understanding of temperature impacts on crop physiology and development. This knowledge will be essential for designing and implementing effective adaptation strategies to sustain agricultural productivity under future climate scenarios. Plant responses to temperature vary not only among species but also across the different stages of their life cycle, particularly in terms of phenological development. Each crop species possesses a characteristic thermal range—defined by minimum, optimum, and maximum temperatures—within which growth and development can occur (Hatfield et al., 2008, 2011). These cardinal temperature values differ between vegetative and reproductive phases. Notably, vegetative processes such as leaf and node appearance rates typically have higher optimum temperatures than those required for reproductive development.

The sensitivity of crops to extreme temperature events is inherently species-specific. For instance, maize (*Zea mays* L.), a warm-season crop, has a much higher temperature tolerance threshold compared to cool-season vegetables such as broccoli (*Brassica oleracea* L.), which exhibits optimal growth around 25°C, versus 38°C for maize. Therefore, what constitutes an "extreme temperature" for one crop may be within the optimal range for another. This variability highlights the importance of interpreting meteorological extremes in the context of a plant's physiological temperature response. In annual (non-perennial) crops, accelerated development under high temperatures leads to shorter life cycles, which can result in reduced plant size, abbreviated reproductive phases, and ultimately lower yield potential. When temperatures rise above or fall below critical thresholds during key developmental stages, such as flowering or grain filling, productivity can be significantly compromised.

Photoperiod-sensitive crops like soybean are especially vulnerable, as temperature fluctuations can disrupt the timing of developmental cues linked to day length, further complicating phenological progression. Across species, elevated temperatures during the reproductive stage are particularly damaging. High temperatures reduce pollen viability, hinder fertilization, and impair grain or fruit formation

(Hatfield et al., 2008, 2011). Chronic exposure to heat stress during pollination and early grain set phases can severely depress yield potential. However, even brief episodes of acute heat stress during reproductive development may be equally or more detrimental. The influence of climate change on crop productivity is most apparent in its impact on yield—a critical metric for both producers and consumers. While alterations in the duration of a crop's life cycle may be tolerated to some extent, yield losses due to thermal stress pose a more significant concern. Yield responses to increasing temperatures are shaped by the crop's intrinsic cardinal temperature thresholds. Consequently, global warming is expected to alter plant developmental patterns and reduce yield potential in temperature-sensitive crops unless adaptive strategies are implemented.

5.4 LOW TEMPERATURE FOR MAIZE

Plant species originating from tropical climates, such as maize (*Zea mays* L.), tend to be particularly vulnerable to chilling stress (Miedema, 1982; Sanghera et al., 2011). Chilling stress is defined as the physiological response to non-freezing, sub-optimal temperatures above 0°C, typically ranging from 0°C to 15°C (Revilla et al., 2005; Zhu et al., 2007; Sanghera et al., 2011; Miura & Furumoto, 2013). While often conflated with freezing stress, the two are physiologically and genetically distinct (Revilla et al., 2005). Chilling stress exerts a direct thermal constraint on cellular processes, whereas freezing stress, occurring at sub-zero temperatures, leads primarily to osmotic damage caused by extracellular ice crystal formation and subsequent cellular dehydration (Hincha & Zuther, 2020). The sensitivity of maize to chilling varies depending on the developmental stage and the plant organ involved. For instance, the root system is particularly susceptible, and any chilling-induced inhibition of root growth can impair water and nutrient uptake later in the plant's life cycle. The most chilling-sensitive stages occur early in development, particularly during seed imbibition and the onset of photosynthesis in germinating seedlings (Revilla et al., 2005). These early disruptions have downstream consequences for plant establishment and productivity.

Experimental studies commonly use temperatures within the 0–15°C range to simulate chilling stress in maize (Hu et al., 2017; Frascaroli & Revilla, 2019). Maize's tropical origin contributes to its general intolerance of cold; however, adaptation to cooler climates has occurred over centuries. For example, temperate varieties such as those belonging to the "Flint" race, widely used in European breeding programs, tend to exhibit greater chilling tolerance compared to tropical or subtropical genotypes (Riva-Roveda et al., 2016). These Flint lines are particularly valuable in northern Europe, where they are incorporated into breeding strategies to enhance cold tolerance. In temperate regions, maize cultivation has expanded steadily over recent decades (Fracheboud & Leipner, 1999; Frascaroli & Revilla, 2019). Early sowing is often employed in these areas to maximize biomass accumulation, minimize exposure to drought and pests, and promote rapid canopy closure, which suppresses weed competition. However, early planting also exposes emerging seedlings to heightened chilling stress and disease pressure, making crop establishment more challenging (Jompuk et al., 2005). In practical agricultural terms, chilling stress in maize is generally considered to begin at temperatures below 10–15°C, with significant physiological impairments evident below 10°C (Hu et al., 2017; Frascaroli & Revilla, 2019). While growth slows markedly below 15°C, the threshold for stress increases to approximately 20°C in more mature plants. Temperatures below 5°C can inflict serious cellular and tissue damage, including injury to seeds and young seedlings (Frascaroli & Revilla, 2019). Germination and photosynthesis are particularly vulnerable, with both processes being severely inhibited below 10°C (Janowiak et al., 2002; Foyer et al., 2002). Despite these challenges, substantial genetic variability exists within maize germplasm for chilling tolerance. This variation provides opportunities for breeding and selecting cold-resilient genotypes. From an agronomic standpoint, maize should ideally be sown when soil temperatures consistently exceed 8°C to minimize the risk of severe chilling stress and ensure successful stand establishment (Sobkowiak et al., 2014; Jończyk et al., 2017).

5.5 LOW-TEMPERATURE IMPACTS

5.5.1 Agronomic

Maize (*Zea mays* L.) is inherently sensitive to cold, which presents a major constraint to its productivity in cool and humid regions such as Central Europe. Exposure to low temperatures in these environments reduces the crop's growth rate while simultaneously prolonging the growth duration. This dual effect weakens seedling vigor early in the season and often results in premature termination of grain filling toward the end of the growth cycle (Thakur et al., 2010; Wijewardana et al., 2016; Hussain et al., 2019). Consequently, maize grown in mountainous and temperate zones frequently exhibits reduced and inconsistent grain yields. Chilling injury, especially during the early seedling or reproductive stages, depends on both the duration and severity of the stress. Low-temperature stress, typically defined as exposure to temperatures below 10°C for an extended period, disrupts normal plant physiological and developmental processes from germination through to reproductive maturity (Farooq et al., 2009). Chen et al. (2012) demonstrated that such stress impairs seedling germination, restricts early growth, and compromises the antioxidant defense system, rendering the plants more vulnerable to oxidative damage.

Cold stress also significantly affects root morphology, photosynthetic efficiency—particularly photosystem II (PSII) performance—chlorophyll content, and leaf area expansion (Hund et al., 2007). A brief chilling episode—such as a drop below 10°C for seven consecutive days during the V6–V9 vegetative stages—can markedly delay the onset of anthesis, which is critical for successful pollination and kernel development. Morphological abnormalities such as malformed tassels have been observed under cold stress, further exacerbating reproductive inefficiency. Given that more than 50% of a maize plant's assimilates are allocated to grain development during the reproductive stage, any temperature stress at this point can have severe consequences for final yield (Hayashi, 2016). Cold stress has also been shown to reduce plant height and total aboveground biomass (Liu et al., 2016). At the cellular level, leaf development is impaired due to prolonged cell cycle duration and a decreased rate of mitosis, collectively resulting in stunted vegetative growth.

5.5.2 Physio-Metabolic

Low temperatures between 8°C and 10°C delay seedling emergence in maize and lead to reductions in root-to-shoot ratio and chlorophyll content during early development (Bano et al., 2015). Chilling temperatures ranging from 4°C to 10°C severely inhibit chlorophyll biosynthesis and suppress PSII activity, thus impairing photosynthetic efficiency (Riva-Roveda et al., 2016). At the cellular level, cold stress damages chloroplast and thylakoid ultrastructure, disrupts enzyme activities, and impairs Calvin cycle function by restricting metabolite transport (Sun et al., 2017). Investigations into cell wall dynamics under chilling stress (12–14°C) have revealed reduced pectin content and diminished pectin methylesterase activity in cold-sensitive maize hybrids, indicating compromised cell wall integrity (Bilska-Kos et al., 2017). Such stress conditions cause various physiological and biochemical dysfunctions, including membrane instability, disrupted enzymatic function, and damage to the photosynthetic apparatus (Liu et al., 2016). At the seedling stage, chilling injury elevates levels of malondialdehyde (MDA) and cell membrane permeability, while simultaneously reducing water content, proline concentration, and chlorophyll levels in leaves (Chen et al., 2012). Furthermore, cold stress restricts the uptake and transport of essential macronutrients such as nitrogen (N), phosphorus (P), potassium (K), calcium (Ca), and magnesium (Mg), leading to nutrient deficiencies in both shoots and roots (Liu et al., 2016). In response, maize activates cold defense pathways, producing signaling molecules like nitric oxide and abscisic acid (Esim et al., 2016). The oxidative stress induced by chilling results from excessive generation of reactive oxygen species (ROS), which cause extensive damage to macromolecules and cell membranes (Apel & Hirt, 2004; Hussain et al., 2018, 2019). To mitigate this, maize plants enhance their antioxidant defenses by increasing levels of enzymes such as superoxide dismutase (SOD), peroxidase (POD), and accumulating proline (Hussain et al., 2019; Hassanuzzaman et al., 2020). Chilling stress during the grain filling phase alters starch biosynthesis, notably by reducing amylose content. This leads to lower starch solubility, reduced swelling power, and increased gelatinization temperature—factors that collectively compromise grain quality (Lu et al., 2014). Moreover, temperatures below 15°C during the reproductive phase suppress key enzymes involved in carbohydrate metabolism, such as sucrose phosphate synthase, phosphoenolpyruvate carboxylase, and sucrose synthase (Yan et al., 2015). These disruptions destabilize the carbon assimilation process, ultimately resulting in suboptimal grain composition and poor kernel texture. In summary, cold stress significantly reduces germination percentage, hampers growth and photosynthesis, and deteriorates both yield and grain quality in maize.

5.6 CONCLUSION

Low-temperature stress represents a significant environmental constraint to maize productivity, especially in regions with temperate or high-altitude climates. Due to its tropical origin, maize is inherently sensitive to chilling and freezing temperatures, particularly during early developmental stages such as germination, seedling emergence, and the transition to reproductive growth. Temperatures below 15°C can impair physiological functions, including photosynthesis, enzyme activity, membrane stability, and hormonal balance, ultimately leading to delayed growth, reduced biomass, and poor grain yield. Understanding the specific temperature thresholds and stage-specific sensitivities is essential for managing maize under low-temperature stress. Considerable genetic variability exists in maize for cold tolerance traits, offering valuable opportunities for breeders to develop resilient hybrids through selection and molecular approaches. Integrating physiological insights, genetic tools, and agronomic practices will be crucial for enhancing maize adaptation to suboptimal temperatures and ensuring sustainable production in the face of climatic uncertainties.

REFERENCES

Apel, K., & Hirt, H. (2004). Reactive oxygen species: metabolism, oxidative stress, and signal transduction. *Annu. Rev. Plant Biol.*, *55*(1), 373–399.

Badu-Apraku, B., Hunter, R. B., & Tollenaar, M. (1983). Effect of temperature during grain filling on whole plant and grain yield in maize (Zea mays L.). *Canadian Journal of Plant Science*, 63(2), 357–363.

Bano, S., Aslam, M., Saleem, M., Basra, S., & Aziz, K. (2015). Evaluation of maize accessions under low temperature stress at early growth stages. *Journal of Animal and Plant Sciences*, 25, 392–400.

Barlow, K. M., Christy, B. P., O'Leary, G. J., Riffkin, P. A., & Nuttall, J. G. (2015). Simulating the impact of extreme heat and frost events on wheat crop production: A review. *Field Crops Research*, 171, 109–119.

Bewley, J. D., Bradford, K., & Hilhorst, H. (2012). *Seeds: physiology of development, germination and dormancy*. Springer Science & Business Media.

Bilska-Kos, A., Solecka, D., Dziewulska, A., Ochodzki, P., Jończyk, M., Bilski, H., & Sowiński, P. (2017). Low temperature caused modifications in the arrangement of cell wall pectins due to changes of osmotic potential of cells of maize leaves (Zea mays L.). *Protoplasma*, 254, 713–724.

Chen, J., Xu, W., Velten, J., Xin, Z., & Stout, J. (2012). Characterization of maize inbred lines for drought and heat tolerance. *Journal of Soil and Water Conservation*, 67, 354–364.

Crafts-Brandner, S. J., & Salvucci, M. E. (2002). Sensitivity of photosynthesis in a C_4 plant, maize, to heat stress. *Plant Physiology*, 129(4), 1773–1780.

Dupuis, L., & Dumas, C. (1990). Influence of temperature stress on maize kernel development. *Plant Physiology and Biochemistry*, 28(4), 459–466.

Esim, N., & Atici, Ö. (2016). Relationships between some endogenous signal compounds and the antioxidantsystem in response to chilling stress in maize (Zea mays L.) seedlings. *Turkish Journal of Botany*, *40*(1), 37–44.

Farooq, M., Aziz, T., Wahid, A., Lee, D.-J., & Siddique, K. H. (2009). Chilling tolerance in maize: Agronomic and physiological approaches. *Crop and Pasture Science*, 60, 501–516.

Foyer, C. H., Vanacker, H., Gomez, L. D., & Harbinson, J. (2002). Regulation of photosynthesis and antioxidant metabolism in maize leaves at optimal and chilling temperatures: Review. *Plant Physiology and Biochemistry*, 40, 659–668.

Foyer, C. H., van Camp, W., & Hanke, G. (2002). The function of ascorbate–glutathione cycle in plants. *Plant Physiology*, 109(4), 1129–1139.

Fracheboud, Y., Haldimann, P., Leipner, J., & Stamp, P. (1999). Chlorophyll fluorescence as a selection tool for cold tolerance of photosynthesis in maize (Zea mays L.). *Journal of Experimental Botany*, 50, 1533–1540.

Fracheboud, Y., & Leipner, J. (1999). The application of chlorophyll fluorescence to study light, temperature, and drought stress in plants. *Journal of Experimental Botany*, 50(336), 1225–1233.

Frascaroli, E., & Revilla, P. (2019). Genomics of cold tolerance in maize. In: Bennetzen J, Flint-Garcia S, Hirsch C, & Tuberosa R, eds. *The maize genome*. Cham: Springer Nature, 287–304.

Greaves, J. A. (1996). Improving suboptimal temperature tolerance in maize—the search for variation. *Journal of Experimental Botany*, 47(3), 307–323.

Hasanuzzaman, M., Bhuyan, M. B., Zulfiqar, F., Raza, A., Mohsin, S. M., Mahmud, J. A., ... & Fotopoulos, V. (2020). Reactive oxygen species and antioxidant defense in plants under abiotic stress: Revisiting the crucial role of a universal defense regulator. *Antioxidants*, *9*(8), 681.

Hatfield, J. L., Boote, K. J., Fay, P., Hahn, L., Izaurralde, R. C., Kimball, B. A., Mader, T., Morgan, J., Ort, D., Polley, W., Thomson, A., & Wolfe, D. (2008). Agriculture In:. The Effects of Climate Change on Agriculture, Land Resources, Water Resources, and Biodiversity in the United States.

Hatfield, J. L., Boote, K. J., Kimball, B. A., Ziska, L. H., Izaurralde, R. C., Ort, D., Thomson, A. M., & Wolfe, D. W. (2011). Climate impacts on agriculture: Implications for crop production. *Agronomy Journal*, 103, 351–370.

Hatfield, J. L., & Prueger, J. H. (2015). Temperature extremes: Effect on plant growth and development. *Weather and Climate Extremes*, 10, 4–10.

Hayashi, T. (2016). Varietal difference in the effects of low temperature on tassel development in hybrid maize. *Plant Production Science*, 19, 230–237.

Herrero, M. P., & Johnson, R. R. (1981). Drought stress and its effects on maize reproductive systems. *Crop Science*, 21(1), 105–110.

Hincha, D. K., & Zuther, E. 2020. Plant cold acclimation and winter survival. In: Hincha DK, & Zuther E, eds. *Plant cold acclimation*. New York: Springer Nature, 1–7.

Hope, H. J., Maamari, R., Séguin, S., Hamilton, R. I., Dwyer, L. M., & White, R. P. (1992). Low temperature emergence potential of short season corn hybrids grown under controlled environment and plot conditions. *Canadian Journal of Plant Science*, 72(1), 83–91.

Hu, G., Li, Z., & Lu, Y. et al. (2017). Genome-wide association study identified multiple genetic loci on chilling resistance during germination in maize. *Scientific Reports*, 7, 10840.

Hund, A., Richner, W., Soldati, A., Fracheboud, Y., & Stamp, P. (2007). Root morphology and photosynthetic performance of maize inbred lines at low temperature. *European Journal of Agronomy*, 27, 52–61.

Huner, N. P. A., Öquist, G., & Sarhan, F. (1998). Energy balance and acclimation to light and cold. *Trends in Plant Science*, 3(6), 224–230.

Hussain, H. A., Hussain, S., Khaliq, A., Ashraf, U., Anjum, S. A., Men, S., & Wang, L. (2018). Chilling and drought stresses in crop plants: Implications, cross talk, and potential management opportunities. *Frontiers in Plant Science*, 9, 393.

Hussain, H. A., Men, S., Hussain, S., Ashraf, U., Zhang, Q., Anjum, S. A., ... & Wang, L. (2019). Individual and concurrent effects of drought and chilling stresses on morpho-physiological characteristics and oxidative metabolism of maize cultivars. *BioRxiv*, 829309.

Hussain, S., Khaliq, A., Ali, B., Hussain, H. A., Qadir, T., & Hussain, S. (2019). Temperature extremes: Impact on rice growth and development. In *Plant abiotic stress tolerance: agronomic, molecular and biotechnological approaches* (pp. 153–171). Cham: Springer International Publishing.

Intergovernmental Panel Climate Change (IPCC). (2007). *Climate change 2007: Impacts, adaptation and vulnerability: Contribution of working group II to the fourth assessment report of the intergovernmental panel on climate change*, Cambridge, UK and New York, NY, Cambridge University Press.

Janowiak, F., Maas, B., & Dörffling, K. (2002). Importance of abscisic acid for chilling tolerance of maize seedlings. *Journal of Plant Physiology*, 159, 635–643.

Jompuk, C., Fracheboud, Y., Stamp, P., & Leipner, J. (2005). Mapping of quantitative trait loci associated with chilling tolerance in maize (Zea mays L.) seedlings grown under field conditions. *Journal of Experimental Botany*, 56, 1153–1163.

Jończyk, M., Sobkowiak, A., Trzcinska-Danielewicz, J., Skoneczny, M., Solecka, D., Fronk, J., & Sowiński, P. (2017). Global analysis of gene expression in maize leaves treated with low temperature. II. Combined effect of severe cold (8°C) and circadian rhythm. *Plant Molecular Biology*, 95, 279–302.

Kumudini, S., Andrade, F. H., Boote, K. J., Brown, G. A., Dzotsi, K. A., Edmeades, G. O., Gocken, C. F., Goodwin, M., Halter, A. L., Hammer, G. L., Hatfield, J. L., Jones, J. W., Kemanian, A. R., Kim, S.-H., Kiniry, J., Lizaso, J. I., Nendel, C., Nielsen, R. L., Parent, B., Stöckle, C. O., Tardieu, F., Thomison, P. R., Timlin, D. J., Vyn, T. J., Wallach, D., Yang, H. S., & Tollenaar, M. (2014). Predicting maize phenology: Inter comparison of functions for developmental response to temperature. *Agronomy Journal*, 106, 2087–2097.

Leipner, J., Fracheboud, Y., & Stamp, P. (2008). Effect of high temperature on photosynthetic acclimation and chilling tolerance in maize. *Journal of Experimental Botany*, 59(15), 4113–4123.

Li, X., Wang, G., Fu, J., Li, L., Jia, G., Ren, L., & Gu, R. (2018). QTL mapping in three connected populations reveals a set of consensus genomic regions for low temperature germination ability in Zea mays L. *Frontiers in Plant Science*, 9, 65.

Liu, Q., Hallerman, E., Peng, Y., & Li, Y. (2016). Development of Bt rice and Bt maize in China and their efficacy in target pest control. *International Journal of Molecular Sciences*, *17*(10), 1561.

Liu, W., Tollenaar, M., Stewart, G., & Deen, W. (2013). Response of maize grain yield to cold stress during vegetative and reproductive stages. *Canadian Journal of Plant Science*, 93(3), 507–518.

Liu, Y., Liang, J., Sun, L., Yang, X., & Li, D. (2016). Group 3 LEA protein, ZmLEA3, is involved in protection from low temperature stress. *Frontiers in Plant Science*, 7, 1011.

Lu, D., Cai, X., Yan, F., Sun, X., Wang, X., & Lu, W. (2014). Effects of high temperature after pollination on physicochemical properties of waxy maize flour during grain development. *Journal of the Science of Food and Agriculture*, 94, 1416–1421.

Lukatkin, A.S. (2003). Contribution of Oxidative Stress to the Development of Cold-Induced Damage to Leaves of Chilling-Sensitive Plants: 3. Injury of Cell Membranes by Chilling Temperatures. *Russian Journal of Plant Physiology*, 50, 243–246. https://doi.org/10.1023/A:1022985500733

Meehl, G. A., Stocker, T. F., Collins, W. D., Gaye, A. J., Gregory, J. M., Kitoh, A., Knutti, R., Murphy, J. M., Noda, A., Raper, S. C. B., Watterson, J. G., Weaver, A. J., & Zhao, Z., 2007. Global climate projections. In: Solomon S., Qin D., Manning M., Chen Z., Marquis M., Averyt K. B., Tignor M., & Miller H. L. eds., *Climate change 2007: The physical science basis. Contribution of Working Group I to the Fourth Assessment Report of the Intergovernmental Panel on Climate Change*. Cambridge: Cambridge University Press.

Miedema, P. (1982). The effects of low temperature on Zea mays. *Advances in Agronomy*, 35, 93–128.

Miura, K., & Furumoto, T. (2013). Cold signaling and cold response in plants. *International Journal of Molecular Sciences*, 14, 5312–5337.

Prasad, P. V. V., Bheemanahalli, R., & Jagadish, S. V. K. (2017). Field crops and the fear of heat stress—opportunities, challenges and future directions. *Field Crops Research*, 200, 114–121.

Rasmussen, A. M., Johansen, I. E., & Andersen, S. B. (1998). Temperature and light quality influence maize seedling development. *Seed Science Research*, 8(2), 151–160.

Revilla, P., Butrón, A., Cartea, M., Malvar, R., & Ordás, A. (2005). Breeding for cold tolerance. In: Ashraf M., & Harris P. J. C., eds. *Abiotic stresses. Plant resistance through breeding and molecular approaches*. Boca Raton, FL: CRC Press, 301–398.

Reynolds, M., Bonnett, D., & Chapman, S. C. (2010). Raising yield potential of wheat: Designing ideotypes, exploiting genetic diversity and optimizing inputs. *Journal of Experimental Botany*, 61(2), 351–362.

Riva-Roveda, L., Escale, B., Giauffret, C., & Périlleux, C. (2016). Maize plants can enter a standby mode to cope with chilling stress. *BMC Plant Biology*, 16, 1–14.

Riva-Roveda, L., Escale, B., Giauffret, C., & Périlleux, C. (2016). Maize plants can enter a standby mode to cope with chilling stress. *BMC Plant Biology*, 16, 212.

Rodríguez, V. M., Romay, M. C., Ordás, A., & Revilla, P. (2010). Evaluation of European maize (Zea mays L.) germplasm under cold conditions. *Genetic Resources and Crop Evolution*, 57, 329–335.

Sanghera, G., Wani, S., Hussain, W., & Singh, N. (2011). Engineering cold stress tolerance in crop plants. *Current Genomics*, 12, 30–43.

Schoper, J. B., Lambert, R. J., Vasilas, B. L., & Westgate, M. E. (1986). Pollen viability, pollen shedding, and combining ability for tassel heat tolerance in maize. *Crop Science*, 26(5), 1029–1033.

Sobkowiak, A., Jończyk, M., Jarochowska, E., Biecek, P., TrzcinskaDanielewicz, J., Leipner, J., Fronk, J., & Sowiński, P. (2014). Genome-wide transcriptomic analysis of response to low temperature reveals candidate genes determining divergent cold-sensitivity of maize inbred lines. *Plant Molecular Biology*, 85, 317–331.

Sun, J., Zheng, T., Yu, J., Wu, T., Wang, X., Chen, G., Tian, Y., Zhang, H., Wang, Y., & Terzaghi, W. (2017). TSV, a putative plastidic oxidoreductase, protects rice chloroplasts from cold stress during development by interacting with plastidic thioredoxin Z. *New Phytologist*, 215, 240–255.

Thakur, P., Kumar, S., Malik, J. A., Berger, J. D., & Nayyar, H. (2010). Cold stress effects on reproductive development in grain crops: An overview. *Environmental and Experimental Botany*, 67, 429–443.

Tollenaar, M., & Lee, E. A. (2002). Yield potential, yield stability and stress tolerance in maize. *Field Crops Research*, 75(2–3), 161–169.

Verheul, M. J., Picatto, C., & Stamp, P. (1996). Growth and development of maize (Zea mays L.) seedlings under chilling conditions in the field. *European Journal of Agronomy*, 5(1–2), 31–43.

Wahid, A., Gelani, S., Ashraf, M., & Foolad, M. R. (2007). Heat tolerance in plants: An overview. *Environmental and Experimental Botany*, 61(3), 199–223.

Wijewardana, C., Henry, W. B., Hock, M. W., & Reddy, K. R. (2016). Growth and physiological trait variation among corn hybrids for cold tolerance. *Canadian Journal of Plant Science*, 96, 639–656.

Yan, G. U., Cao, M. K., Zhang, Y. Q., Sun, Y., Wen-He, H. U., & Chun-Sheng, W. U. (2015). Effects of low temperature on photosynthetic fluorescence and enzyme activity of carbon metabolism of maize at booting stage. *Journal of Plant Physiology*, 51, 941–948.

Zhu, J., Dong, C., & Zhu, J. (2007). Interplay between cold-responsive gene regulation, metabolism and RNA processing during plant cold acclimation. *Current Opinion in Plant Biology*, 10, 290–295.

6 Adjustments of Maize to Low-Temperature (LT) Stress

6.1 INTRODUCTION

Maize (*Zea mays* L.) is a thermophilic C_4 crop naturally adapted to high solar radiation, elevated daytime temperatures, and moderate photoperiod sensitivity. Temperature is a major determinant of maize productivity, profoundly influencing growth dynamics, phenology, and yield potential. Optimal growth and physiological performance occur at 28–30°C, supporting efficient enzymatic activity, photosynthesis, and metabolic coordination (Kumar et al., 2021). Deviations from this range, particularly exposure to low temperatures (LTs) (~10°C), suppress enzyme activity, reduce chlorophyll synthesis, impair water and nutrient uptake, and lead to stunted growth, chlorosis, and leaf whitening. Prolonged chilling further compromises seed germination, seedling establishment, and vegetative vigor, with the seedling stage, spanning germination to the three-leaf stage, being especially sensitive to LT stress (Xuhui et al., 2022). LT exposure also affects reproductive development by delaying tassel emergence, silk elongation, and gametogenesis, thereby extending the anthesis–silking interval (ASI) and reducing kernel set (Larden & Triboi-Blondel, 1994; Angadi et al., 2000). Anatomically, chilling reduces mesophyll and bundle sheath cell size, restricts intercellular air spaces, and impairs CO_2 diffusion (Janowiak & Dorffling, 1996). Physiologically, LT stress induces oxidative damage, membrane destabilization, enzyme inactivation, and cellular leakage, while osmoregulatory solutes such as soluble sugars and proline accumulate to maintain hydration and metabolic stability (Cai et al., 2014). Hormonal responses, including elevated abscisic acid (ABA) and salicylic acid (SA) with decreased gibberellic acid (GA), further modulate adaptive responses under LT stress (Pal et al., 2020). At the molecular level, LT stress tolerance in maize is governed by a complex interplay of genomic, transcriptomic, and proteomic networks. Quantitative genetic analyses reveal that additive and non-additive gene actions, maternal effects, and gene–environment interactions regulate LT-responsive traits, providing targets for breeding stress-resilient genotypes (Turk et al., 2019; Neta et al., 2020). Transcriptome studies highlight dynamic expression of cold-responsive (COR) genes, including DREB1/CBF transcription factors and regulatory networks associated with photosynthesis, stress signaling, and metabolic adaptation (Li et al., 2016; Waititu et al., 2021; Proteomic investigations further identify LT-responsive proteins that contribute to signaling, metabolism, and cellular defense (Xu et al., 2013; Chen et al., 2015). Post-transcriptional and

DOI: 10.1201/9781003634027-6

post-translational modifications (PTMs), including alternative splicing, phosphorylation, ubiquitination, and SUMOylation, fine-tune the stability and activity of key regulators such as ICE1, CBFs, and COR proteins, integrating environmental cues with growth and stress adaptation.

This chapter provides a comprehensive overview of maize responses to LT stress, encompassing morphological, anatomical, physiological, and molecular mechanisms, with a focus on genomics, transcriptomics, proteomics, and post-transcriptional and post-translational regulatory pathways.

6.2 MORPHOLOGICAL AND ANATOMICAL BEHAVIOR

Maize (*Zea mays* L.) is a thermophilic C_4 crop naturally adapted to environments characterized by intense solar radiation, elevated daytime temperatures, and relatively low sensitivity to photoperiod. Among various climatic determinants influencing maize productivity, temperature exerts a profound regulatory effect on growth dynamics, phenological development, and yield potential. The species exhibits consistently high thermal requirements throughout its life cycle, although the optimal temperature range varies among developmental stages. Vigorous growth and efficient physiological performance are typically achieved within a temperature range of 28–30°C (Kumar et al., 2021). Within these limits, enzymatic reactions, photosynthetic activity, and metabolic coordination function optimally. Departures from this optimal range, either toward lower or higher temperatures, disrupt essential physiological and biochemical processes. At suboptimal temperatures (around 10°C), maize exhibits suppressed enzyme activity, diminished chlorophyll synthesis, and impaired water and nutrient absorption. These effects manifest as stunted growth, chlorosis, and leaf whitening. Prolonged exposure to such chilling conditions further hinders seed germination, seedling establishment, and vegetative vigor. Conversely, temperatures above 32°C accelerate leaf senescence, reduce pollen viability, and compromise kernel set, indicating that both thermal extremes adversely impact crop performance (Cofas 2018). The base temperature for maize development is approximately 8°C across most phenophases, though early emergence requires slightly higher thermal accumulation. Du Plessis (2003) emphasized that maintaining a minimum air temperature of 10–15°C for about seven consecutive days ensures uniform germination, whereas temperatures below 10°C nearly arrest growth. From an agronomic standpoint, optimizing sowing dates and thermal regimes is vital to prevent the most heat- and water-sensitive phase—flowering—from coinciding with episodes of midseason heat or drought stress. Such synchronization minimizes reproductive failures and promotes efficient grain filling. LT stress, generally defined as exposure to temperatures below 10°C for a prolonged duration, disrupts physiological processes across all growth stages—from germination and early seedling development to anthesis and grain filling (Farooq et al., 2009). The magnitude of injury is strongly influenced by the duration, timing, and intensity of exposure, as well as the intrinsic tolerance capacity of the genotype. Despite its overall susceptibility to cold, maize demonstrates certain morphological and anatomical adjustments that mitigate LT-induced injury. Morphologically, tolerant genotypes tend to exhibit thicker leaves, shortened internodes, and compact canopies that collectively reduce transpirational water loss and maintain internal water balance (Hund et al., 2008). The root system also undergoes adaptive modification, characterized by shorter yet thicker roots that enhance mechanical stability and increase the root-to-shoot ratio, thereby improving water and nutrient uptake efficiency under chilling stress (Hund et al., 2008).

6.2.1 Morphological Changes

6.2.1.1 Germination and Seedling Growth

Maize (*Zea mays* L.) is inherently a warm-season crop and is generally not cultivated in regions where the mean daily temperature falls below 19°C or the average temperature during the summer months remains under 23°C (Du Plessis, 2003). Although maize seeds can germinate at a minimum temperature of about 10°C, optimal and more uniform germination occurs at soil temperatures ranging between 16°C and 18°C. At approximately 20°C, seedlings typically emerge within five to six days, whereas temperatures around 32°C are considered critical, leading to detrimental effects on yield (Du Plessis, 2003). Impaired seedling growth under stress conditions has been attributed to inefficient mobilization of seed reserves and reduced protein synthesis, as observed in maize seedlings exposed to elevated temperatures (Riley, 1981). However, LTs are equally detrimental. Chen et al. (2012) demonstrated that cold stress significantly limits maize germination and seedling growth by destabilizing the antioxidant defense mechanism. Similarly, temperatures around 8–10°C markedly delay seedling emergence and reduce root/shoot ratio and chlorophyll content during early growth (Bano et al., 2015). Exposure to temperatures between 4°C and 10°C can further inhibit chlorophyll biosynthesis and severely impair photosystem II (PSII) activity (Riva et al., 2016). The seedling stage, extending from the completion of germination to the three-leaf stage, represents a critical developmental window in maize, wherein plant morphology and potential yield are determined. This stage is particularly sensitive to LT stress (Xuhui et al., 2022). The extent of morphological establishment and biomass accumulation during this phase largely governs subsequent plant vigor and productivity. When maize seedlings experience chilling stress during this period, the accumulation of organic matter and subsequent developmental processes are significantly restricted, ultimately reducing yield potential (Zhao et al., 2019; Yi et al., 2021).

6.2.1.2 Vegetative Growth and Leaf Development

Although maize (*Zea mays* L.), as a C_4 crop, possesses inherently high photosynthetic efficiency, this potential is substantially constrained under LT conditions. Exposure to chilling temperatures causes notable reductions in plant height, leaf number, leaf area, and overall fresh biomass (Strigens et al., 2013). Under severe temperature drops, yield losses can exceed 15% (Wang et al., 2008). LT stress also affects root system traits by reducing dry and fresh weight, altering branching patterns, and modifying root morphology (Kaspar & Bland, 1992). Cold stress inhibits both cell division and cell elongation, leading to the formation of shorter, thicker leaves with restricted laminar expansion. Impaired chlorophyll biosynthesis and decreased photosynthetic activity often give the leaves a dark green or purplish hue due to anthocyanin accumulation (Hund et al., 2008). In addition, leaf rolling and premature senescence are frequently observed as adaptive mechanisms to reduce transpirational water loss. These morphological restrictions collectively limit canopy development and light interception, thereby decreasing total biomass accumulation. Temperature exerts a strong regulatory influence on leaf initiation and expansion and, more broadly, governs vegetative and reproductive shoot development (Farooq et al., 2009). Chilling conditions during early spring planting have been associated with poor seedling establishment and weak stand development (Ahmad et al., 2012). The rate of leaf area expansion is highly temperature dependent (Lee et al., 2002). Under controlled chilling stress (15/3°C; 16 h photoperiod), the rate of leaf emergence in maize is nearly three times slower compared to plants grown under optimal temperatures (25/15°C; 16 h photoperiod). Similarly, under field conditions, insufficient root development, closely linked to low soil temperature, further restricts leaf growth (Farooq et al., 2009). At the morphological level, cold stress induces visible symptoms such as leaf yellowing, wilting, and reduced leaf unfolding efficiency (Cofer et al., 2018). These responses, which manifest early during the seedling phase, represent the first indicators of LT-induced injury. Altogether, LT stress delays vegetative growth, suppresses leaf expansion, and reduces canopy photosynthetic capacity, ultimately translating into diminished biomass accumulation and yield potential.

6.2.1.3 Root System Architecture

Roots are among the plant organs most sensitive to chilling. LT stress suppresses meristematic activity at root tips, leading to reduced primary root elongation, fewer lateral roots, and a decreased total root surface area (Farooq et al., 2009). Consequently, the overall root system becomes compact, limiting its efficiency in water and nutrient absorption. Chilling also disturbs the balance of key phytohormones, such as auxins and cytokinins, which further impairs root branching and elongation, affecting both plant anchorage and structural stability. Cold stress diminishes root hydraulic conductance (Aroca et al., 2001; Melkonian et al., 2004), negatively influencing water relations and nutrient uptake, thereby restricting overall plant growth potential (Farooq et al., 2009). Additionally, LTs inhibit primary root growth and reduce the branching angles between primary and lateral roots, further compromising the spatial expansion of the root system (Nagel et al., 2009; Frey et al., 2020). At the cellular level, maize epidermal cells are particularly vulnerable to cold, exhibiting the highest number of differentially expressed genes (DEGs) in response to chilling (Zhou et al., 2024). Root hairs—tubular extensions of epidermal cells—play a crucial role in enhancing root surface area, improving water and nutrient uptake, and providing additional mechanical support (Peterson & Farquhar, 1996). Therefore, root hairs are essential for the adaptation and resilience of maize roots under LT stress.

6.2.1.4 Reproductive Growth

Exposure to LTs during the reproductive phase of maize delays tassel emergence and silk elongation, leading to an expanded ASI. Cold stress reduces pollen viability and germination, resulting in poor fertilization and kernel abortion (Shen et al., 2017).

The combined effects of delayed floral development, impaired gametophyte function, and reduced assimilate supply often lead to shorter ears, decreased kernel set, and significant yield losses. Structural and functional abnormalities in reproductive organs under chilling conditions can cause fertilization failure or premature seed abortion. During such stress, reproductive tissues, including the tapetum, style, and endosperm, experience nutrient deficiencies due to restricted source-to-sink solute mobilization (Nayyar et al., 2005). Following the vegetative phase, reproductive development is initiated through the apical floral transition, during which the indeterminate pattern of leaves and internodes is altered to form a determinate floral structure (Meyerowitz et al., 1989; Chimonidou-Pavlidou, 2004). Flowering time in maize is primarily governed by photoperiod and temperature cues (Summerfield & Wien, 1980; Roberts et al., 1985; Lawn et al., 1995). Suboptimal temperatures can sometimes induce early flowering through vernalization effects on actively dividing cells, typically within meristematic zones (Chouard, 1960; Wellensiek, 1964). Ovules in maize are particularly sensitive to chilling stress from anthesis through early grain development (Westgate & Boyer, 1985). Overall, LT affects all stages of the reproductive cycle, including gametogenesis, pollination, fertilization, and embryogenesis (Larden & Triboi-Blondel, 1994; Angadi et al., 2000). Chilling prolongs the cell cycle and reduces the rate of cell division in susceptible species such as maize (Francis & Barlow, 1988; Rymen et al., 2007). Regulation of the cell cycle is a critical determinant of organ growth under adverse conditions. Temperature influences enzyme kinetics involved in cell cycle control, such as cyclin-dependent kinase CDKA, thereby modulating the growth rate of reproductive and vegetative tissues (Granier et al., 2000). For instance, low night temperatures were shown to inhibit maize leaf growth by prolonging meristematic cell cycles (Rymen et al., 2007). During grain development, chilling stress is expected to reduce cell division activity, thereby decreasing sink strength in the endosperm and limiting assimilate import. Sink strength, defined as the product of seed weight and relative growth rate, governs the efficiency of grain filling (Bell & Incoll, 1990). After fertilization, hydrolytic changes in seeds establish a translocation gradient, marking the seed as a "sink" (Venkateswarlu et al., 1986). Environmental constraints during seed development strongly influence both seed quantity and quality (Zhao et al., 2008). Kernel weight is determined by both the rate and duration of grain filling, with rate playing a more decisive role (Daynard et al., 1971; Badu-Apraku et al., 1983). Chilling reduces both the rate and duration of grain filling, thereby limiting the accumulation of storage proteins, minerals, and amino acids (Nayyar et al., 2007). Cold stress can also indirectly reduce source strength by diminishing leaf area, as reduced cell division and expansion under chilling lower leaf size (Ben-Haj-Salah & Tardieu, 1995; Ercoli et al., 2004). In maize, Rymen et al. (2007) reported a 20% reduction in leaf size following exposure to LTs.

6.2.2 Anatomical Changes

The growth of leaves, as well as vegetative and reproductive shoots, is strongly influenced by temperature (Bhandari, 2018). LTs reduce the rate of cell division and elongation, which can result in smaller leaf area and limited seedling growth even when temperatures later rise. Chilling also slows the rate of leaf emergence, potentially reducing the total number of leaves produced (Fryer et al., 1998; Ying et al., 2000). At the cellular level, LTs can damage plant cell membranes, increase membrane permeability, and cause leakage of electrolytes and intracellular substances (Orvar et al., 2000; Hussain et al., 2018). In studies of plant cold tolerance, anatomical traits serve as key indicators for assessing stress responses (Soltabayeva et al., 2021; Yavas et al., 2024). Under LT stress, leaf thickness, spongy mesophyll thickness, and bundle sheath thickness are often reduced (Cao et al., 2014; Momayyezi et al., 2022). Conversely, plants can enhance cold resistance by increasing the thickness of the cuticle, epidermis, and leaf blades (Rape, 1999; Equiza et al., 2001; Cai & Song, 2001). Chloroplast structure is particularly vulnerable to chilling. LTs partially destroy chloroplasts, reduce their numbers, and distort membrane organization. Cellular cavitation, plasma membrane damage, and cytoplasmic disintegration are commonly observed in cold-stressed leaves (Zhang & Zhu, 2012; Wen et al., 2014; Venzhik et al., 2014). Notably, in cold-tolerant maize genotypes, leaf and mesophyll cell layer thickness increase under chilling, whereas cold-sensitive genotypes exhibit the opposite pattern (Bilska-Kos et al., 2018; Venzhik & Moshkov, 2023). Anatomical observations further reveal that mesophyll and bundle sheath cell size is reduced under LT stress, limiting intercellular air spaces and CO_2 diffusion (Janowiak & Dorffling, 1996). Chilling also alters cell membrane structure, increases permeability, causes cytoplasmic extravasation, and reduces hydraulic conductivity. In cold-sensitive genotypes, membrane damage impairs recovery by disrupting root hydraulic function and overall cellular integrity (Aroca et al., 2005). Lipidomics and transcriptomics studies have shown increases in phosphatidic acid and digalactosyl diacylglycerol, decreases in phosphatidylcholine and monogalactosyl diacylglycerol, and active regulation of membrane lipids under LT conditions (Gu et al., 2017). Compared to sensitive inbred lines, tolerant maize genotypes accumulate more unsaturated fatty acids and secondary metabolites, maintaining membrane stability and protecting photosystem structure (Li et al., 2019). In senescing chloroplasts of chilling-sensitive lines, oxygen evolution is impaired and free fatty acids accumulate, whereas tolerant lines show reduced oxygen release without significant accumulation of free fatty acids or accelerated leaf senescence (Saczynska et al., 1993). Additionally, LTs alter cell wall biochemical components closely associated with membrane integrity (Bilska-Kos et al., 2018).

6.3 PHYSIOLOGICAL RESPONSES

LT exerts a pronounced influence on maize growth and development, particularly during the germination, seedling, and flowering stages. Physiologically, exposure to LT stress disrupts critical early processes such as seed germination, seedling vigor, overall germination rate and potential, and root system architecture. Consequently, during the subsequent seedling phase, cold stress hampers the degree and uniformity of germination, reduces seedling establishment and growth potential, and often results in dwarf plants with poorly developed stems and roots. At the reproductive stage, suboptimal temperatures interfere with ear differentiation during tasseling, impair pollination efficiency at flowering, and reduce grain-filling rate and dry matter accumulation, ultimately delaying grain maturity and reducing yield potential. On the molecular and cytological levels, LT stress primarily manifests through alterations in membrane integrity and fluidity, disturbances in photosynthetic and metabolic physiology, and modifications in osmoregulatory and antioxidant enzyme activities, as well as hormonal balance. These physiological and biochemical perturbations form the foundation of maize's chilling stress responses and have therefore been the focus of extensive research aimed at elucidating the mechanisms underlying cold tolerance. Each of these aspects is discussed in detail in the subsequent sections.

6.3.1 Antioxidant Defense System

LT stress induces oxidative damage in maize cells, primarily through the overaccumulation of reactive oxygen species (ROS) such as hydrogen peroxide (H_2O_2), superoxide anion (O_2^-), hydroxyl radicals, and malondialdehyde (MDA) (Prasad et al., 1994; Aroca et al., 2005; Hussain et al., 2020). To mitigate this oxidative stress, plants activate a complex antioxidative defense system involving the upregulation of total glutathione levels and glutathione reductase activity (Kocsy et al., 2002), as well as enhanced expression of catalase (CAT) and esterase enzymes (Neta et al., 2020). Oxidative stress induced by LT causes severe cellular injury, including leakage of intracellular organic acids, enzyme inactivation, and damage to proteins and nucleic acids. Such disruptions create a conducive environment for fungal and bacterial proliferation, ultimately resulting in seed decay, powdering, and rotting (Cao et al., 2019). Various seed priming and exogenous treatments have been explored to alleviate cold-induced oxidative damage. Seed priming with melatonin or chitosan has been shown to enhance antioxidant enzyme activities, thereby improving germination potential and rate under chilling conditions (Guan et al., 2009; Cao et al., 2019). Similarly, seeds coated with SA exhibit elevated activity of protective enzymes, higher germination rates, and improved root and shoot growth under LT stress (Guan et al., 2015). Supplementation of the growth medium with silica also enhances superoxide dismutase (SOD) activity and boosts antioxidant content, strengthening the plant's defense system (Moradtalab et al., 2018). Exogenous application of diethyl aminoethyl hexanoate has been reported to suppress LT-mediated O_2^- generation and excessive H_2O_2 accumulation, while maintaining the net photosynthetic rate, thereby improving the cold tolerance of maize seedlings (Zhang et al., 2020). LT-tolerant maize inbred line CO 328 exhibited significantly higher SOD activity compared to the sensitive line CO 316 (Pinhero et al., 1997). Cold-acclimated maize seedlings also display elevated CAT, glutathione reductase, and guaiacol peroxidase (POD) activities during both chilling and recovery phases, relative to non-acclimated seedlings (Prasad, 1997). Additionally, pretreatment of 3-day-old maize seedlings with H_2O_2 or menadione at ambient temperature has been found to induce cold tolerance through increased CAT and guaiacol POD activities (Prasad et al., 1994). Enhanced antioxidant defense capacity is often correlated with improved morphological traits under LT stress. For instance, the maize hybrid XD889 demonstrated higher activities of SOD, POD, CAT, and glutathione reductase, along with greater root length, biomass, volume, and surface area under LT conditions compared to hybrid XD319 and inbred lines Yu13 and Yu37 (Hussain et al., 2020). At the molecular level, genetic regulation plays a vital role in modulating oxidative stress responses. Overexpression of the ZmASR1 (ABA-stress-ripening) gene in maize led to reduced MDA accumulation and enhanced SOD and POD activities, resulting in improved cold tolerance (Li et al., 2018). Similarly, overexpression of the R2R3-MYB transcription factor *ZmMYB31* in *Arabidopsis thaliana* suppressed LT-induced ROS accumulation and elevated SOD and ascorbate peroxidase (APX) activities, enhancing cold tolerance in transgenic lines (Li et al., 2019). Moreover, *ZmMKK1* overexpression in tobacco increased antioxidant enzyme activities, promoted accumulation of osmoprotectants, and upregulated ROS- and stress-responsive genes, collectively conferring higher cold tolerance in transgenic plants (Cai et al., 2014).

6.3.2 Accumulation of Osmolytes/Biochemicals

Osmoregulatory substances play a crucial role in protecting maize cells against LT stress by maintaining cellular hydration and stabilizing metabolic processes. Under chilling conditions, the accumulation of compatible solutes such as soluble sugars and proline increases significantly, helping to prevent cellular water loss and sustain normal physiological activities (Cai et al., 2014; Guan et al., 2019). In LT-sensitive genotypes, sucrose accumulates rapidly following exposure to cold, leading to alterations in cellular osmotic potential and a reduction in both cell wall pectin content and pectin esterase activity (Bilska-Kos et al., 2017). Conversely, cold-tolerant maize genotypes exhibit a substantial accumulation of cryoprotectants, including free proline and soluble sugars, while maintaining lower levels of oxidative stress markers such as MDA. This balanced response contributes to a stronger antioxidant defense system and enhanced tolerance to chilling stress (Duran Garzon et al., 2020). Chilling-tolerant genotypes also display higher glucose-6-phosphate dehydrogenase (G6PDH) activity and an increased sucrose-to-starch ratio compared to sensitive genotypes, underscoring the importance of starch–sucrose metabolism as a key determinant of the LT response (Duran Garzon et al., 2020). Furthermore, transcriptional regulation contributes significantly to osmoprotection; for instance, *ZmDREB1A* enhances raffinose biosynthesis by activating the expression of *ZmRAF* (raffinose synthase), a critical enzyme in the raffinose synthesis pathway, thereby improving LT tolerance in maize (Han et al., 2020).

6.3.3 Role of Plant Hormones

Plant hormone dynamics undergo significant modulation under LT stress, reflecting their crucial role in mediating maize's adaptive responses. Typically, ABA and SA levels increase, while GA levels decline during exposure to chilling conditions (Pal et al., 2020). The exogenous application of these hormones has been demonstrated to markedly influence maize cold tolerance. Pre-application of ABA prior to cold exposure induces alterations in protein synthesis, enabling maize suspension-cultured cells to acquire enhanced tolerance to LTs (Xin & Li, 1993). Similarly, maize seedlings pretreated with ABA exhibit elevated antioxidant enzyme activities and greater resilience to cold stress (Prasad et al., 1994). Transcriptomic analyses further reveal that the ability of maize to withstand LT stress is strongly associated with the upregulation of ABA biosynthesis in roots, leaves, and crowns (Xue et al., 2021). Exogenous SA application also plays a pivotal role in mitigating chilling injury. Pretreating maize radicles or leaves with SA prior to cold stress reduces leaf electrolyte leakage and enhances the activities of glutathione reductase and guaiacol POD, thereby improving chilling tolerance (Kang & Saltveit, 2002; Waqas et al., 2017). Moreover, the combined application of SA and hydrogen peroxide (H_2O_2) synergistically enhances antioxidant enzyme expression, GA biosynthesis, and GA–ABA signaling, leading to accelerated germination, improved seed vigor, and enhanced seedling growth under LT conditions (Li et al., 2017). Treatment with 24-epibrassinolide (EBR) has also been reported to significantly improve cold tolerance in maize seedlings by increasing tissue water content, maintaining membrane integrity, and elevating chlorophyll, soluble sugar, and protein levels (Singh et al., 2012). In contrast, overexpression of the *ZmPgb1.1* (Phytoglobin) gene in maize suppresses nitric oxide and brassinolide biosynthesis, as well as brassinolide signaling, thereby mitigating the adverse effects of LT stress (Mira et al., 2021).

6.3.4 Photosynthesis and Photoinhibition

LT exerts a profound impact on the photosynthetic performance of maize, disrupting multiple components of the photosynthetic machinery. Chilling stress significantly reduces the maximum catalytic activity of photosynthetic enzymes (Kingston-Smith & Foyer, 2000), PSII efficiency (Ortiz-Lopez et al., 1990; Govindachary et al., 2004; Bilska-Kos et al., 2018), and leaf chlorophyll and protein contents, as well as CO_2 assimilation rates (Kingston-Smith & Foyer, 2000; Riva-Roveda et al., 2016). These physiological impairments collectively constrain carbon fixation and energy conversion, leading to reduced plant growth and productivity under cold stress. Experimental studies involving cold acclimation and light intensity treatments have revealed that exposure to high-intensity light during acclimation enhances the adaptive capacity of maize to subsequent LT stress. Such preconditioning improves chilling tolerance by elevating soluble sugar content and promoting photoprotective mechanisms (Szalai et al., 2018). Structural and functional analyses of chloroplasts further demonstrate genotype-specific responses to chilling. The LT-sensitive inbred line CM 109 shows greater impairment in chloroplast photosynthetic efficiency and ultrastructure than the tolerant inbred line KW 1074, indicating inherent genotypic differences in photosynthetic resilience (Sowinski et al., 2005). In maize leaves, photosynthetic reduction under chilling stress has been linked to plasmodesmatal closure between mesophyll and bundle sheath cells, as well as between bundle sheath and vascular parenchyma cells, thereby restricting intercellular communication and metabolite transport (Bilska & Sowinski, 2010). During germination, maize seeds exposed to LTs sustain growth partly through regulatory adjustments in photosynthetic pathways; however, chilling-sensitive genotypes exhibit more pronounced damage to the photosynthetic system compared to tolerant ones (Meng et al., 2022). Following cold acclimation (14/12°C), chilling-tolerant maize plants display significantly lower injury to the photosynthetic apparatus upon subsequent exposure to severe cold stress (8/6°C), underscoring the protective effect of acclimation mechanisms. Advances in molecular studies have identified several genetic determinants that enhance photosynthetic stability under LT. For instance, overexpression of RAF1–LSSS (*Rubisco assembly factor 1–Rubisco large and small subunits*) in maize improves photochemical quenching before and after chilling exposure, thereby accelerating post-stress recovery (Salesse-Smith et al., 2020). Similarly, overexpression of the R2R3–MYB transcription factor *ZmMYB–IF35* in *Arabidopsis thaliana* alleviates PSII photoinhibition and elevates antioxidant enzyme activity, conferring enhanced tolerance to LT stress (Meng & Sui, 2019).

6.4 MOLECULAR RESPONSES

6.4.1 Genomics

Cold tolerance in maize is a quantitative agronomic trait governed by the coordinated action of multiple genes and their interactions with environmental factors (Turk et al., 2019). In recent years, the use of advanced genetic and genomic approaches has greatly contributed to dissecting the complex genetic architecture underlying this trait. Quantitative genetic analyses have revealed that the expression of cold tolerance in maize is influenced by a diverse set of genetic components, including heterosis, general and specific combining abilities (GCA and SCA), reciprocal maternal and non-maternal effects, additive and non-additive gene actions, genotypic variation, growth stage, and environmental interactions. Early diallel studies were instrumental in elucidating these relationships. For example, Hodges et al. (1997a,b) investigated cold tolerance at both the germination and seedling stages using twelve hybrids derived from a complete diallel of four inbred lines. Their findings demonstrated significant variations in seed germination rate, the activities of CAT and APX in leaves at the three-leaf stage, as well as differences in GCA and SCA for dry matter accumulation at the four-leaf stage. Similarly, Revilla et al. (2000) reported that cold tolerance in maize is influenced by additive, dominant, and maternal genetic effects, highlighting the multifactorial nature of this trait. Further supporting evidence was provided by Yan et al. (2017), who observed significant heterosis for cold tolerance during the seedling stage, emphasizing the potential of hybrid vigor in enhancing chilling resilience. Expanding on this, Neta et al. (2020) evaluated three cold-tolerant and three cold-sensitive inbred lines through a partial diallel cross, assessing heterosis, GCA, SCA, reciprocal effects, and maternal influences, alongside the expression of genes encoding antioxidant enzymes such as CAT, APX, and SOD. Their results confirmed the presence of heterosis and reciprocal effects under cold stress, indicating that non-additive gene action plays a predominant role in determining maize cold tolerance. Interestingly, the genetic control of cold tolerance appears to vary with developmental stage. According to Frascaroli and Revilla (2018), traits expressed during germination, emergence, and early seedling growth are likely regulated by distinct genetic models, suggesting that cold tolerance mechanisms are stage-specific rather than universally controlled across the plant's life cycle. Adding to the complexity, gene–environment interactions have been reported to significantly influence the expression of cold tolerance traits, further complicating the genetic landscape of this response (Presterl et al., 2007). Collectively, these findings underscore that the genetics of cold tolerance in maize is highly intricate, involving polygenic control, environmental modulation, and developmental specificity. Nevertheless, with the advent of genomic tools and high-throughput molecular analyses, substantial progress has been achieved in recent years in identifying quantitative trait loci (QTLs) and candidate genes associated with cold tolerance. The integration of these findings offers new opportunities for a comprehensive molecular understanding of maize cold tolerance and its application in breeding programs for stress-resilient genotypes.

6.4.1.1 QTL Mapping

QTL mapping serves as a robust strategy to identify and manipulate loci underlying complex agronomic traits, including cold tolerance, in crop plants. In maize, this approach has been extensively applied to dissect the genetic basis of cold tolerance using various segregating populations such as $F_{2:3}$ families, recombinant inbred lines (RILs), and backcross (BC) populations. Most of these populations were generated through crosses between cold-tolerant and cold-sensitive inbred lines to maximize phenotypic variation in response to LT. Diverse physiological and morphological parameters—including seed germination rate, root architecture, and photosynthetic efficiency—were evaluated under both optimal and LT regimes to reveal loci associated with cold responsiveness. Several reliable molecular markers have been identified and linked to QTLs or genes regulating cold tolerance, offering valuable resources for breeding cold-tolerant maize genotypes. For instance, Fracheboud et al. (2002) conducted QTL analysis using 233 maize RILs and identified 18 QTLs significantly correlated with photosynthetic performance at LTs. The major QTL associated with leaf development was also found to influence pigment composition, suggesting pleiotropic effects. Similarly, Fracheboud et al. (2004) utilized an $F_{2:3}$ population derived from ETH-DH7 × ETH-DL3 and detected 19 QTLs governing chlorophyll fluorescence under chilling stress, with the principal locus located on chromosome 6—later confirmed by Jompuk et al. (2005) as a major QTL influencing both chlorophyll fluorescence and pigment content. Other studies have highlighted loci affecting root and shoot growth under cold conditions. Hund et al. (2004) mapped a dominant QTL on chromosome 5 controlling root and shoot development in an $F_{2:3}$ population derived from Lo964 × Lo1016. Presterl et al. (2007) identified 18 QTLs for traits such as leaf chlorosis, anthocyanin accumulation, and frost injury using 720 doubled haploid (DH) lines. Similarly, Rodríguez et al. (2008) detected two QTLs significantly correlated with leaf color at LTs on chromosomes 3 and 6, explaining 14.2% of phenotypic and 28.2% of genetic variation in an IBM (intermated B73 × Mo17) population. Subsequent studies refined these findings further. Guerra-Peraza et al. (2012) identified a major photosynthesis-related QTL on chromosome 5 using the IBM302 population, with the favorable allele contributed by Mo17. Rodríguez et al. (2013, 2014) identified genomic regions associated with cold-induced albinism and photosystem stability in a cross between A661 (susceptible) and EP42 (tolerant), pinpointing a major QTL on chromosome 2 explaining 14% of phenotypic variation and three meta-QTLs regulating seedling development under cold stress.

Hu et al. (2016) mapped six QTLs each for LT germination rate and primary root length in 243 IBM Syn4 RILs, with four overlapping genomic regions, while Yan et al. (2017) reported seven QTLs controlling four cold-related traits in an $F_{2:3}$ population (K932 × Mei C), each explaining 10.55–25.29% of phenotypic variation. To improve resolution, advanced mapping populations have been developed. Shi et al. (2016) employed two connected RIL populations sharing a common parent, identifying 26 QTLs for seed vigor under cold stress, later consolidated into five meta-QTLs, with two major loci on chromosomes 2 and 9 ($R^2 > 10\%$). Likewise, Li et al. (2018) used three connected $F_{2:3}$ populations from cold-tolerant and cold-sensitive lines to detect 43 QTLs and three meta-QTL regions associated with seed germination ability under chilling conditions. Yi et al. (2020) leveraged a 406-line multi-parent advanced generation intercross (MAGIC) population to identify 148 QTLs encompassing 858 SNPs, with most loci localized to bin 10.04. In recent years, QTL mapping has been integrated with genome-wide association studies (GWASs) and transcriptomic analyses to uncover candidate genes with greater accuracy. Goering et al. (2021) combined QTL mapping and transcriptome profiling in a subset of 97 maize lines to identify two major COR QTLs and 13 candidate genes implicated in chlorophyll concentration and tissue integrity under chilling stress. Jin et al. (2021) further demonstrated the efficacy of joint QTL–GWAS analysis by identifying 12 QTLs associated with POD activity in an $F_{2:3}$ population (W10 × W72) and four SNPs linked to the same trait in a natural population of 80 inbred lines. Through integrative analysis, Zm00001d002729 was pinpointed as a key cold tolerance gene whose overexpression enhanced maize seedling survival by elevating POD activity and reducing MDA accumulation and ion leakage.

Collectively, these studies reveal that the genomic regions governing cold tolerance in maize are widely distributed across nearly all chromosomes. However, despite the identification of numerous QTLs, few have been fine-mapped or functionally characterized. This limitation largely stems from the immense labor and population size required for detailed genetic dissection. Nonetheless, with the advancement of joint mapping approaches and integrative genomic analyses, the identification of causal genes—such as Zm00001d002729 (Jin et al., 2021)—demonstrates the promising potential of modern QTL-based methodologies for elucidating cold tolerance mechanisms in maize.

6.4.1.2 GWAS

In addition to biparental segregating populations traditionally used in QTL mapping, natural populations comprising hundreds of genetically diverse maize inbred lines have been extensively employed to investigate the genetic architecture of cold tolerance through GWASs (Frascaroli & Revilla, 2018). Although GWAS and QTL analyses often focus on similar COR physiological and developmental traits—such as seed germination rate, root length, and leaf pigmentation—GWAS offers superior mapping resolution and precision due to the greater allelic diversity and historical recombination present in the mapping populations (Lipka et al., 2015). The first GWAS aimed at dissecting cold tolerance traits in maize was conducted by Strigens et al. (2013) using a diverse germplasm panel of 375 inbred genotypes genotyped with 56K single nucleotide polymorphisms (SNPs). The study identified 19 significant association signals, explaining 5.7–52.5% of the phenotypic variation for cold-related parameters, including early growth and chlorophyll fluorescence.

Following this, Huang et al. (2013) evaluated 125 maize inbred lines for 10 cold tolerance-related traits at the germination and seedling stages, detecting 43 SNPs associated with cold tolerance and predicting 40 candidate genes based on 31 of these SNPs. To further explore genetic variation, Yan et al. (2017) developed a test-cross association mapping population by crossing the cytoplasmic male sterile line *S-Mo17* with 338 diverse inbred lines. GWAS performed on this population revealed 19 SNPs significantly associated with cold tolerance-related traits. Similarly, Hu et al. (2017) performed GWAS on 282 maize inbred lines under both normal and LT conditions, identifying 17 SNPs linked to cold tolerance, seven of which were located within candidate genes. Subsequent studies expanded the scope of GWAS for cold stress response in maize. Zhang et al. (2020) used 222 inbred lines to identify 30 SNPs associated with cold tolerance during seed germination, leading to the discovery of 14 candidate genes, whose expression profiles were validated under chilling stress. In a later study, Zhang et al. (2021) analyzed 300 inbred lines for germination-related traits at LTs and detected 15 significant SNPs, with three genomic loci repeatedly associated with multiple cold tolerance traits. In one of the most comprehensive studies to date, Yi et al. (2021) evaluated 836 maize inbred lines and uncovered 187 significant SNPs, which were consolidated into 159 genomic regions influencing seed emergence and early growth traits under LT stress. Despite the enhanced resolution and broader genetic coverage offered by GWAS, certain limitations and controversies remain. Notably, many cold-associated SNP markers are located within non-coding genomic regions, suggesting regulatory rather than coding functions. However, determining which specific genes these SNPs regulate remains challenging. The extensive structural variation among maize genomes adds further complexity—since genetic loci identified in the reference *B73* genome may not fully represent the genomic architecture of other inbred lines. Consequently, some unpredicted or lineage-specific genes may lie in close proximity to cold-associated SNPs, potentially leading to incomplete or inaccurate gene–trait associations. Overall, GWAS represents a valuable complement to traditional QTL mapping by providing high-resolution insights into the polygenic basis of cold tolerance. Yet, the integration of GWAS with other genomic and transcriptomic approaches remains essential for validating causal genes and understanding the functional networks underlying cold stress adaptation in maize.

6.4.2 Transcriptomics

Transcriptome analyses have emerged as indispensable tools for deciphering the molecular mechanisms underlying cold stress tolerance in maize. By enabling genome-wide detection of gene expression changes, these approaches provide crucial insights into the signaling networks, regulatory genes, and pathways activated during cold stress. A wide array of transcriptomic studies, based on RNA sequencing (RNA-seq), microarrays, and miRNA profiling, have been conducted to identify COR genes and elucidate their functional roles in stress adaptation. Using RNA-seq, Li et al. (2016) compared the transcriptomic profiles of a freezing-tolerant (KR701) and a freezing-sensitive (Hei8834) maize line at the seedling stage before and after cold exposure. They identified 948 DEGs, with gene ontology (GO) analysis revealing an enrichment in "binding functions," "protein kinase activity," and "peptidase activity," suggesting their involvement in cold signaling and protein turnover. Similarly, Li et al. (2019b) analyzed cold-tolerant (M54) and cold-sensitive (753F) maize inbred lines under cold stress and reported a greater number of DEGs in M54 after 4 and 24 hours of treatment, implying a more robust and dynamic transcriptional response in tolerant genotypes. Expanding upon these findings, Li et al. (2020) profiled the transcriptome of *B73* seedlings under different temperature regimes (4°C, 10°C, and 16°C) and detected 5,358, 5,485, and 5,312 DEGs, respectively, including significant upregulation of five key genes such as *ZmDERB1*. A population-based transcriptomic study by Frey et al. (2020) examined 21 DH lines from the flint landrace "Petkuser Ferdinand rot," comprising 11 cold-resistant and 10 cold-sensitive lines. Differential expression analysis identified 148 DEGs related to cold treatment, 3,254 DEGs linked to cold tolerance, and 563 DEGs associated with growth rate at LT, emphasizing the genetic diversity in transcriptional responses to cold. The study by Zhang et al. (2020) further validated candidate genes identified via GWAS by RNA-seq and discovered 10 DEGs located within linkage disequilibrium regions, two of which were implicated in cold signal transduction and membrane fluidity regulation. Li et al. (2021) investigated transcriptomic changes in seeds of three sweet corn near-isogenic lines (NILs) and their parents under cold conditions, identifying 20 DEGs highly correlated with germination performance; notably, a UDP-glucosyltransferase gene was hypothesized to play a central role in cold-induced germination. A large-scale comparative RNA-seq study by Waititu et al. (2021) analyzed seedlings of 24 cold-tolerant and 22 cold-sensitive inbred lines, identifying 2,237 DEGs, including 147 transcription factors (TFs) from 32 families such as *MYB*, *ERF*, *NAC*, *WRKY*, *bHLH*, *MIKC MADS*, and *C2H2*. Similarly, Yu et al. (2021) conducted a time-course analysis of leaf transcriptomes in two contrasting inbred lines and revealed that cold tolerance in line *B144* was achieved through active regulation of stomatal movement and protection of PSII from photooxidative damage by upregulating *D1* protein–encoding genes. Conversely, the sensitive line *Q319* failed to modulate stomatal closure under cold stress, highlighting key physiological divergence between tolerant and susceptible lines.

In addition to RNA-seq, several earlier transcriptomic approaches have also contributed valuable insights. Rodríguez et al. (2013) employed microarray hybridization to study chlorophyll-deficient (albino) and chlorophyll-containing leaf sections of the cold-sensitive line *A661*, identifying 1,002 differentially expressed transcripts across 23 functional categories, including genes involved in tetrapyrrole biosynthesis and photosynthesis. Similarly, Di Fenza et al. (2017) used microarray analysis across four varieties—two cold-tolerant and two cold-sensitive—and reported 64 DEGs in tolerant lines, with negligible changes in sensitive genotypes. Complementary studies using cDNA-AFLP identified three COR genes—*ZmMAPKKK*, *ZmCLCD*, and *ZmRLK*—that may participate in stress perception and signaling (Yang et al., 2011). Regulatory non-coding RNAs also play crucial roles in maize cold response. Aydinoglu (2020) profiled the miRNA landscape (miRNome) of the hybrid ADA313 under cold stress and found 24, 6, and 20 differentially expressed miRNAs in the meristem, elongation, and mature zones, respectively. These findings underscore the spatial regulation of miRNAs in modulating tissue-specific cold responses. Further integrating genomic and computational approaches, Zhou et al. (2022) examined transcriptomic changes in seedlings of *B73*, *Mo17*, and *W22* inbred lines, along with their three F_1 hybrids (*B73×Mo17*, *W22×B73*, and *W22×Mo17*), under cold and heat stress. Their analysis not only identified extensive stress-related DEGs but also differentiated cis- and trans-regulatory mechanisms of expression divergence. Importantly, they demonstrated how sequence variation in cis-elements influences gene responsiveness to stress and proposed advanced machine learning models (e.g., convolutional neural networks) to predict plant stress responses based on regulatory architecture.

Collectively, transcriptomic studies have identified hundreds to thousands of DEGs across experiments, yet the overlap among them remains remarkably low. Sowinski et al. (2020), after surveying 13 independent transcriptomic studies conducted on maize seedlings (V2–V5 stages), reported that fewer than 500 DEGs were common to more than one study, while most DEGs were study-specific. Of these, four studies employed moderate LT stress, while nine used severe LT stress, and the DEGs showed fundamentally distinct expression patterns between these two regimes. This variability underscores that gene expression is highly influenced by genotype, tissue type, developmental stage, experimental design, and stress intensity. Despite this heterogeneity, the conserved DEGs shared across multiple studies represent promising targets for functional validation and could serve as key molecular indicators for breeding cold-tolerant maize varieties. Future integration of transcriptomics with proteomic, metabolomic, and epigenomic data will be essential to fully unravel the complex gene networks governing maize adaptation to LTs.

6.4.3 Proteomics

Understanding plant responses to cold stress requires an in-depth examination of protein turnover and abundance under LT conditions. LT stress triggers the accumulation of various proteins in both tolerant and sensitive plants, a phenomenon observed across several species (Koehler et al., 2012; Grimaud et al., 2013; Xu et al., 2013; Chen et al., 2015). Such analyses of protein dynamics under LT stress help uncover key regulatory pathways and candidate genes that can be targeted to enhance crop tolerance. Proteomics, therefore, serves as an indispensable tool for identifying stress-inducible proteins and elucidating the physiological and molecular mechanisms that govern LT stress adaptation (Zhang et al., 2017b). Protein profiling in plants can be accomplished through various gel-based and gel-free techniques. Conventional gel-based approaches such as two-dimensional gel electrophoresis (2-DE) and two-dimensional differential gel electrophoresis (2D-DIGE) allow visualization of changes in protein abundance. More advanced gel-free quantitative approaches, including isotope-coded affinity tags (ICAT), isobaric tags for relative and absolute quantitation (iTRAQ), and stable isotope labeling by amino acids in cell culture (SILAC), offer higher sensitivity and throughput (Yates et al., 2009). Over the past decade, these methodologies have been increasingly employed to decipher proteomic responses of cereals, particularly maize, to LT stress. Comprehensive proteomic investigations in maize have provided significant insights into the mechanisms of LT tolerance. For instance, Yu et al. (2025) applied tandem mass tag (TMT)-labeled quantitative proteomics combined with data-independent acquisition (DIA) and liquid chromatography–tandem mass spectrometry (LC–MS/MS) to study dynamic proteome alterations in two maize inbred lines (LT-tolerant B144 and LT-sensitive Q319) subjected to 5°C treatment at the three-leaf stage. Out of 4367 non-redundant proteins identified, distinct regulatory patterns were observed between the two lines. B144 exhibited exclusive upregulation under LT stress (6 DEPs at 24 h and 16 DEPs at 48 h), while Q319 displayed mixed regulation (9 DEPs at 24 h, including 6 upregulated and 3 downregulated; and 21 DEPs at 48 h, including 19 upregulated and 2 downregulated). These findings highlight genotype-specific proteomic remodeling under LT stress and underscore the dynamic molecular adjustments that differentiate tolerant from sensitive maize lines. Early molecular responses to cold shock in maize have also been explored through proteome and phosphoproteome profiling. Xing et al. (2022) reported that maize seedlings exposed to 4°C for 30 min and 2 h showed rapid and complex alterations in both protein abundance and phosphorylation patterns. Quantitative analysis identified over 8500 proteins and 19,300 phosphopeptides, of which approximately 660 and 620, respectively, were responsive to cold. Interestingly, only 65 proteins were shared between these datasets, indicating distinct regulatory layers at the protein and PTM levels. Functional enrichment analyses revealed that early cold responses involve pathways associated with photosynthetic light reactions, spliceosome activity, endocytosis, and defense responses, resembling similar mechanisms reported in *Arabidopsis*. A total of 32 photosynthesis-related proteins were downregulated at the abundance level, while 48 spliceosome proteins exhibited altered phosphorylation. Furthermore, 31 kinases and 33 transcription factors were identified as COR, suggesting rapid activation of signaling cascades that fine-tune transcriptional and post-transcriptional responses to temperature decline. This work provides an extensive proteome–phosphoproteome landscape for maize, deepening our understanding of cold signaling in C_4 plants. Ramazan et al. (2023) reported a comparative proteomic investigation to decipher the molecular mechanisms underlying LT stress tolerance in maize. In this study, two contrasting maize genotypes, Gurez local, a cold-tolerant variety from the Kashmir Himalaya, and GM6, a tropical genotype, were analyzed to understand their differential protein expression patterns under LT stress. Leaf tissues at the three-leaf stage were subjected to chilling stress (6°C for 12 h), and the total proteins were resolved through two-dimensional gel electrophoresis (2D-PAGE) followed by MALDI-TOF mass spectrometric identification. The analysis revealed 19 differentially regulated proteins (DRPs) in Gurez local and 10 DRPs in GM6. Interestingly, three novel LT-responsive proteins—threonine dehydratase biosynthetic chloroplastic, thylakoidal processing peptidase 1 chloroplastic, and nodulin-like protein—were identified exclusively in Gurez local. These proteins had not been previously reported in relation to abiotic stress responses, highlighting their potential significance in maize cold tolerance. The higher abundance and diversity of LT-responsive proteins in Gurez local reflected its superior adaptive capacity and more efficient stress perception mechanisms compared to the tropical genotype. Overall, Ramazan et al. (2023) demonstrated that Gurez local exhibits a more dynamic and targeted proteomic response to

LT stress, suggesting that its superior tolerance is linked to early activation of specific protein networks and signaling pathways. The identification of novel LT-inducible proteins provides valuable molecular targets for future functional validation and marker-assisted breeding to develop LT-tolerant maize cultivars. In another iTRAQ-based study, Wang et al. (2016) characterized the differential abundance protein species (DAPS) in leaves of the cold-tolerant maize inbred line W9816 exposed to chilling (4°C) and control (25°C) conditions at the three-leaf stage. A total of 173 DAPS were identified, with 159 annotated in 38 GO functional groups, 108 classified into 20 clusters of orthologous groups, and 99 enriched in KEGG pathways. Antioxidant assays validated the proteomic data, confirming that the adaptive response of maize seedlings to chilling stress involves the alleviation of photodamage, enhanced glycolytic energy production, increased abundance of stress-responsive proteins, and improved ROS scavenging capacity. These results further emphasized the importance of post-transcriptional regulation and PTMs in modulating maize adaptation to LTs. Melatonin (MEL; *N*-acetyl-5-methoxytryptamine), a multifunctional signaling molecule with antioxidant and growth-regulating properties, also plays a crucial role in plant defense against abiotic stress. Kołodziejczyk et al. (2016) examined the effects of melatonin-based seed priming on maize (*Zea mays* L.) proteome under chilling stress. Seeds were hydroprimed with water or melatonin solutions (50 and 500 μM) and subsequently subjected to 5°C stress during germination. Proteome maps generated through two-dimensional electrophoresis and analyzed by mass spectrometry revealed several melatonin-associated proteins differentially expressed under chilling conditions. These proteins were primarily involved in antioxidant defense, energy metabolism, and protein stabilization, demonstrating that melatonin enrichment enhances the stress tolerance capacity of maize seeds during germination under suboptimal temperatures. Further insights into the hormonal regulation of cold response were provided by Zhou et al. (2022), who investigated proteomic changes alongside physiological and phytohormone analyses in two contrasting maize inbred lines—KR701 (tolerant) and Hei8834 (sensitive)—at different seedling stages under chilling stress. The study revealed that the ABA signaling pathway plays a central role in conferring cold tolerance in KR701, particularly at the first-leaf stage. Several signaling proteins and stress-responsive pathways were catalogued and analyzed, providing a valuable resource for future research aimed at elucidating the molecular mechanisms of cold adaptation and for guiding maize breeding programs toward enhanced chilling tolerance.

Collectively, these proteomic investigations reveal that maize responds to LT stress through intricate and multilayered regulatory mechanisms involving changes in photosynthetic activity, energy metabolism, antioxidant defense, phosphorylation signaling, and hormonal balance. The integration of quantitative proteomics with functional genomics and physiological analyses offers a comprehensive framework for identifying candidate proteins and pathways that underpin cold tolerance. Such insights not only advance our understanding of the molecular basis of stress adaptation but also pave the way for developing cold-resilient maize cultivars through targeted breeding and biotechnological interventions.

6.4.4 Post-Transcriptional and Post-Translational Modifications

6.4.4.1 The Post-Transcriptional Modifications

Cold acclimation in plants involves a highly coordinated cascade of transcriptional and post-transcriptional regulatory mechanisms that fine-tune gene expression in response to temperature fluctuations. At the transcriptional level, COR genes often contain *dehydration-responsive element (DRE)/C-repeat (CRT)* cis-elements, characterized by the conserved core motif CCGAC, which serve as binding sites for *C-repeat Binding Factors* (CBFs) or *DREB1* transcription factors (Yamaguchi-Shinozaki & Shinozaki, 1994; Stockinger et al., 1997). The CBF-dependent module represents one of the best-characterized cold regulatory pathways. In *Arabidopsis thaliana*, cold stress rapidly induces the expression of *CBF1*, *CBF2*, and *CBF3*, whose encoded proteins subsequently activate *COLD-REGULATED (COR)* genes, thereby enhancing freezing tolerance (Jaglo-Ottosen et al., 1998; Liu et al., 1998). Functional studies have demonstrated that these CBFs play conserved and crucial roles in mediating cold tolerance across multiple plant species (Shi et al., 2018; Zhang et al., 2019; Zeng et al., 2021). However, constitutive overexpression of CBF genes, while improving cold tolerance, often results in reduced biomass accumulation (Jaglo-Ottosen et al., 1998), suggesting the need for fine-tuned regulation of this pathway. CBFs function as master regulators of *COR* gene expression but also influence signaling pathways linked to phytohormones, light, and circadian rhythms (Ding et al., 2020). Their promoters contain multiple cis-elements that integrate endogenous and environmental signals. Among the upstream regulators, INDUCER OF CBF EXPRESSION 1 (ICE1) is one of the most well-characterized transcription factors controlling cold-mediated induction of CBFs (Chinnusamy et al., 2003). The ICE–CBF–COR signaling module appears to be evolutionarily conserved, with homologs identified in rice and maize contributing to LT tolerance.

In maize, *Zm*ICE1 plays a pivotal role in regulating cold response through both transcriptional and metabolic reprogramming. Under cold stress, DNA demethylation occurs over the *ZmDREB1* promoter region bound by ZmICE1, enabling its activation (Hu et al., 2011). *Zm*ICE1 directly induces *ZmDREB1* expression and thereby enhances LT tolerance (Jiang et al., 2022). In addition, *Zm*ICE1 reprograms the maize amino acid metabolome by repressing *ASPARAGINE SYNTHETASE (ZmAS)* expression, which limits the conversion of glutamine or aspartic acid into asparagine or glutamic acid, consequently reducing mitochondrial ROS production. The lowered ROS levels prevent suppression of *ZmDREB1s*, reinforcing the activation of downstream COR genes (Jiang et al., 2022). Further, *ZmMYB39*, a positive regulator of maize LT tolerance, binds directly to the *ZmICE1* promoter to activate its expression, strengthening the ICE–CBF–COR regulatory loop in maize. Together, these findings underscore the conservation and functional importance of ICE–CBF–COR signaling in plant cold acclimation.

Additional transcriptional regulators, including CALMODULIN-BINDING TRANSCRIPTION ACTIVATORs (CAMTAs), also contribute to CBF induction. In *Arabidopsis*, CAMTA3 and CAMTA5 regulate *CBF1* expression during rapid, but not gradual, temperature decreases (Doherty et al., 2009; Kidokoro et al., 2017).

Since CAMTAs are closely associated with Ca^{2+} signaling—a key early event in cold perception—their activation appears essential for full induction of *COR* genes (Knight et al., 1996). Moreover, repressor-type transcription factors fine-tune this system to maintain homeostasis during prolonged cold exposure. *Arabidopsis MYB15* and *maize bZIP68* directly repress cold-induced expression of *CBF* genes (Agarwal et al., 2006; Li et al., 2022). In *Arabidopsis*, MYB15 degradation by PUB25 and PUB26 E3 ligases during early cold stress and its subsequent reaccumulation at later stages ensure dynamic modulation of CBF expression (Wang et al., 2019). Similarly, in maize, the stability and transcriptional activity of *Zm*bZIP68 are regulated by MPK8-mediated phosphorylation, providing another layer of post-translational control (Li et al., 2022).

While transcriptional regulation of cold responses in maize has been extensively studied, post-transcriptional regulation—involving RNA processing, stability, transport, and translation—also plays a vital role in cold acclimation. Components such as splicing factors, nucleoporins, and RNA helicases are increasingly recognized for their roles in modulating mRNA maturation and export under LT conditions. However, the specific molecular interactions and signaling dynamics of these post-transcriptional regulators remain largely unexplored in maize. Further research into these aspects will be crucial to comprehensively understand how maize integrates transcriptional and post-transcriptional networks to achieve LT tolerance.

6.4.4.2 The Post-Translational Modifications

PTMs play a pivotal role in regulating LT stress responses in plants by modulating the activity, conformation, localization, and stability of key regulatory proteins. Several genes interact through PTMs with the ICE–CBF–COR signaling module, thereby determining the extent of LT tolerance. Among the various types of PTMs, phosphorylation, ubiquitination, and SUMO conjugation are the most significant processes that orchestrate LT stress responses in plants (Sheremet et al., 2012). Under LT stress, the Inducer of CBF Expression 1 (ICE1) protein undergoes multiple PTMs that alter its stability and consequently affect the transcription of CBF and other downstream COR genes.

6.4.4.3 Phosphorylation Modifications

Phosphorylation, a reversible protein modification mediated by kinases and phosphatases, is a key determinant of plant cold adaptation (Ding et al., 2015). The serine/threonine (Ser/Thr) protein kinase OST1 (Open Stomata 1)/SnRK2.6/SRK2E, which functions in the ABA signaling pathway, phosphorylates ICE1 to enhance its transcriptional stability and cold-induced expression (Chen et al., 2021). Furthermore, OST1 prevents ICE1 degradation by interfering with the E3 ubiquitin ligase HOS1, thereby promoting the sustained activation of CBF genes and improving cold tolerance (Msanne et al., 2011).

The Mitogen-Activated Protein Kinase (MAPK) cascade also contributes significantly to cold signal transduction. This cascade typically involves three sequentially acting kinases—MAPKKK, MAPKK, and MAPK—which together transmit extracellular stimuli such as hormones, cytokines, and stress signals to intracellular effectors (Li et al., 2017; Huang et al., 2023). In the COR MAPK pathway, the MEKK1–MKK2–MPK4–MPK6 cascade is central. Upon LT stress, MEKK1 expression is upregulated, activating downstream MKK2, which phosphorylates and activates MPK4. This signaling arm positively regulates cold tolerance. In contrast, MPK3 and MPK6, which are also activated under cold conditions, phosphorylate ICE1 at Ser94, Thr366, and Ser403, leading to its degradation and thereby negatively regulating CBF expression (Wang et al., 2023). Thus, the MEKK1–MKK2–MPK4 cascade promotes cold tolerance, whereas the MPK3/MPK6-mediated pathway serves as a negative feedback mechanism.

6.4.4.4 Ubiquitination and SUMOylation Modifications

Ubiquitination is another crucial PTM that governs the turnover of COR proteins. It involves a sequential enzymatic process mediated by ubiquitin-activating (E1), conjugating (E2), and ligating (E3) enzymes, where the E3 ubiquitin ligase confers substrate specificity (Yang et al., 2021; Tang et al., 2024). In Arabidopsis, HOS1 (High Expression of Osmotically Responsive Gene 1) encodes a RING finger E3 ubiquitin ligase that specifically targets ICE1 for ubiquitin-mediated degradation, thereby attenuating CBF expression and reducing cold tolerance (Chaikam & Karlson, 2010). The *hos1* mutant exhibits a superinduction of CBFs and COR genes under cold stress, highlighting the importance of controlled proteolysis in fine-tuning the COR transcriptome (Lee et al., 2001; Dong et al., 2006). The nuclear localization of HOS1 is enhanced by cold, further facilitating ICE1 ubiquitination. In wild-type plants, polyubiquitination and subsequent proteolysis of ICE1 occur within 12 hours of LT stress, whereas in *hos1* mutants, ICE1 stability is retained, leading to enhanced freezing tolerance (Dong et al., 2006).

In contrast to ubiquitination, SUMOylation (the conjugation of Small Ubiquitin-like Modifier, SUMO, to target proteins) generally stabilizes regulatory proteins and protects them from proteasomal degradation (Ulrich, 2005; Miura et al., 2007). SUMOylation is catalyzed through a cascade involving E1-activating, E2-conjugating, and E3-ligating enzymes, while SUMO proteases (SENPs) remove SUMO groups to maintain homeostasis (Sayed et al., 2024; Imaduwage & Hewadikaram, 2024). The SUMO E3 ligase SIZ1 has been shown to enhance LT tolerance by SUMO-conjugating ICE1 at the K393 residue, thereby preventing its ubiquitination and degradation (Shi et al., 2012). Arabidopsis *siz1* mutants display hypersensitivity to chilling and freezing stress, accompanied by reduced CBF and COR gene expression but elevated levels of AtMYB15, a negative regulator of CBF transcription (Miura et al., 2007). Transgenic plants overexpressing wild-type ICE1 exhibit increased freezing tolerance, whereas those expressing a non-SUMOylatable ICE1 (K393R) mutant remain sensitive to LT, confirming that SIZ1-mediated SUMOylation of ICE1 is essential for its stability and full transcriptional activation of CBFs during cold acclimation. Collectively, these findings reveal that the opposing actions of ubiquitination (via HOS1) and SUMOylation (via SIZ1) on ICE1 act as a molecular rheostat that modulates CBF expression and fine-tunes plant adaptation to LT stress.

Although much of the mechanistic understanding of PTMs under LT stress comes from *Arabidopsis* and other model plants, maize also exhibits distinct post-translational responses to chilling. Notably, the resistance of maize plants to

LT stress has been linked to the appearance of a novel chlorophyll *a/b*-binding protein in the thylakoid membrane following chilling treatment under light. This cold-induced protein was identified as a phosphorylated form of the minor chlorophyll *a/b* protein CP29, rather than a product of a cold-regulated gene or unprocessed precursor (Bergantino et al., 1995). The CP29 kinase responsible for this phosphorylation showed unique characteristics compared to other thylakoid kinases and was found to be regulated by the redox state of a quinone site. Importantly, maize plants lacking this phosphorylation activity displayed enhanced sensitivity to LT-induced photoinhibition, highlighting that post-translational phosphorylation of thylakoid proteins such as CP29 plays a vital protective role during cold stress acclimation in maize. Despite emerging insights into phosphorylation and other PTMs under LT stress, the molecular landscape of PTMs in maize remains largely unexplored. Future studies should focus on integrating phosphoproteomics, ubiquitinomics, and SUMOylomics to unravel genotype-specific PTM networks associated with LT tolerance. Identifying maize-specific E3 ligases, kinases, and SUMO ligases involved in regulating ICE–CBF–COR signaling could provide valuable targets for molecular breeding. Moreover, cross-talk among different PTMs, for instance, the interplay between phosphorylation and ubiquitination in regulating photosystem stability, warrants deeper investigation. Advanced proteomic tools combined with CRISPR-Cas9-mediated functional validation of PTM-related genes can accelerate the development of maize varieties with enhanced LT resilience, offering a promising direction for future cold-tolerance research in maize.

REFERENCES

Agarwal, M., Hao, Y., Kapoor, A., Dong, C. H., & Fujii, H. et al. (2006). A R2R3 type MYB transcription factor is involved in the cold regulation of CBF genes and in acquired freezing tolerance. *Journal of Biological Chemistry*, 281, 37636–37645.

Ahmad, I., Ahmad, T. K. A., Basra, S. M., Hasnain, Z., & Ali, A. (2012). Effect of seed priming with ascorbic acid, salicylic acid and hydrogen peroxide on emergence, vigor and antioxidant activities of maize. *African Journal of Biotechnology*, 11, 1127–1137.

Angadi, S. V., Cutforth, H. W., Miller, P. R., McConkey, B. G., Entz, M. H., Brandt, S. A., & Volkmar, K. M. (2000). Response of three Brassica species to high temperature stress during reproductive growth. *Canadian Journal of Plant Science*, 80(4), 693–701.

Aroca, R., Amodeo, G., Fernandez-Illescas, S., Herman, E. M., Chaumont, F., & Chrispeels, M. J. (2005). The role of aquaporins and membrane damage in chilling and hydrogen peroxide induced changes in the hydraulic conductance of maize roots. *Plant Physiology*, 137, 341–353.

Aroca, R., Tognoni, F., Irigoyen, J. J., Sánchez-Díaz, M., & Pardossi, A. (2001). Different root low temperature response of two maize genotypes differing in chilling sensitivity. *Plant Physiology and Biochemistry*, 39(12), 1067–1073.

Aydinoglu, F. (2020). Elucidating the regulatory roles of microRNAs in maize (Zea mays L.) leaf growth response to chilling stress. *Planta*, 251, 38. doi: 10.1007/s00425-019-03331-y

Badu-Apraku, Á, Hunter, R. B., & Tollenaar, M. (1983). Effect of temperature during grain filling on whole plant and grain yield in maize (Zea mays L.). *Canadian Journal of Plant Science*, 63(2), 357–363.

Bano, S., Aslam, M., Saleem, M., Basra, S., & Aziz, K. (2015). Evaluation of maize accessions under low temperature stress at early growth stages. *Journal of Animal and Plant Sciences*, 25, 392–400.

Bell, C. J., & Incoll, L. D. (1990). The redistribution of assimilate in field-grown winter wheat. *Journal of Experimental Botany*, 41(8), 949–960.

Ben-Haj-Salah, H., & Tardieu, F. (1995). Temperature affects expansion rate of maize leaves without change in spatial distribution of cell length (analysis of the coordination between cell division and cell expansion). *Plant Physiology*, 109(3), 861–870.

Bergantino, E., Dainese, P., Cerovic, Z., Sechi, S., & Bassi, R. (1995). A post-translational modification of the photosystem II subunit CP29 protects maize from cold stress. *Journal of Biological Chemistry*, 270(15), 8474–8481.

Bhandari, K. (2018). Chilling stress: How it affects the plants and its alleviation strategies. *International Journal of Pharmaceutical Sciences and Research*, 9(6), 2197–2200.

Bilska, A., & Sowinski, P. (2010). Closure of plasmodesmata in maize (Zea mays) at low temperature: A new mechanism for inhibition of photosynthesis. *Annals of Botany*, 106, 675–686.

Bilska-Kos, A., Panek, P., Szulc-Głaz, A., Ochodzki, P., Cisło, A., & Zebrowski, J. (2018). Chilling-induced physiological, anatomical and biochemical responses in the leaves of miscanthus× giganteus and maize (Zea mays L.). *Journal of Plant Physiology*, 228, 178–188.

Bilska-Kos, A., Solecka, D., Dziewulska, A., Ochodzki, P., Jonczyk, M., Bilski, H., & Sowinski, P. (2017). Low temperature caused modifications in the arrangement of cell wall pectins due to changes of osmotic potential of cells of maize leaves (Zea mays L.). *Protoplasma*, 254, 713–724.

Cai, G., Wang, G., Wang, L., Pan, J., Liu, Y., & Li, D. (2014). Zmmkk1, a novel group a mitogen-activated protein kinase gene in maize, conferred chilling stress tolerance and was involved in pathogen defense in transgenic tobacco. *Plant Science*, 214, 57–73.

Cai, Y. L., & Song, Y. C. (2001). Adaptive ecology of lianas in Tiantong evergreen broad-leaved forest, Zhejiang, China I. Leaf anatomical characters. *Chinese Journal of Plant Ecology*, 25(1), 90.

Cao, H. X., Huang, H. J., Lei, X. T., Zhang, D. P., Zhang, R. L., & Sun, C. X. (2014). The effect of different low temperature treatment on the anatomical structure of oil palm leaves. *Chinese Journal of Tropical Crops*, 35(3), 454–459.

Cao, Q., Li, G., Cui, Z., Yang, F., Jiang, X., Diallo, L., & Kong, F. (2019). Seed priming with melatonin improves the seed germination of waxy maize under chilling stress via promoting the antioxidant system and starch metabolism. *Scientific Reports*, 9, 15044.

Chaikam, V., & Karlson, D. T. (2010). Response and transcriptional regulation of rice SUMOylation system during development and stress conditions. *BMB Reports*, 43, 103–109.

Chen, J., Han, G., Shang, C., Li, J., Zhang, H., Liu, F., Wang, J., Liu, H., & Zhang, Y. (2015). Proteomic analyses reveal differences in cold acclimation mechanisms in freezing-tolerant and freezing-sensitive cultivars of alfalfa. *Frontiers in Plant Science*, 6, 105. doi: 10.3389/fpls.2015.00105

Chen, J., Xu, W., Velten, J., Xin, Z., & Stout, J. (2012). Characterization of maize inbred lines for drought and heat tolerance. *Journal of Soil and Water Conservation*, 67, 354–364.

Chen, X., Ding, Y., Yang, Y., Song, C., Wang, B., Yang, S., Guo, Y., & Gong, Z. (2021). Protein kinases in plant responses to drought, salt, and cold stress. *Journal of Integrative Plant Biology*, 63, 53–78.

Chimonidou-Pavlidou, D. (2004). Malformation of roses due to drought stress. *Scientia Horticulturae*, 99(1), 79–87.

Chinnusamy, V., Ohta, M., Kanrar, S., Lee, B.-H., & Hong, X. et al. (2003). ICE1: A regulator of cold-induced transcriptome and freezing tolerance in Arabidopsis. *Genes & Development*, 17, 1043–1054.

Chouard, P. (1960). Vernalization and its relations to dormancy. *Annual Review of Plant Physiology*, 11(1), 191–238.

Cofas, E. (2018). The dynamics of maize production in the climate factors variability conditions. In *Agrarian Economy and Rural Development-Realities and Perspectives for Romania. 9th Edition of the International Symposium, November 2018, Bucharest* (pp. 239–245). Bucharest: The Research Institute for Agricultural Economy and Rural Development (ICEADR).

Cofer, T. M., Engelberth, M., & Engelberth, J. (2018). Green leaf volatiles protect maize (Zea mays) seedlings against damage from cold stress. *Plant, Cell & Environment*, 41(7), 1673–1682. doi: 10.1111/pce.13204

Daynard, T. B., Tanner, J. W., & Duncan, W. G. (1971). Duration of the grain filling period and its relation to grain yield in corn, Zea mays L. 1. *Crop Science*, 11(1), 45–48.

Di Fenza, M., Hogg, B., Grant, J., & Barth, S. (2017). Transcriptomic response of maize primary roots to low temperatures at seedling emergence. *PeerJ*, 5, e2839. doi: 10.7717/peerj.2839

Ding, Y., Li, H., Zhang, X., Xie, Q., Gong, Z., & Yang, S. (2015). OST1 kinase modulates freezing tolerance by enhancing ICE1 stability in *Arabidopsis. Developmental Cell*, 32, 278–289.

Ding, Y., Shi, Y., & Yang, S. (2020). Molecular regulation of plant responses to environmental temperatures. *Molecular Plant*, 13, 544–564.

Doherty, C. J., Van Buskirk, H. A., Myers, S. J., & Thomashow, M. F. (2009). Roles for Arabidopsis CAMTA transcription factors in cold-regulated gene expression and freezing tolerance. *Plant Cell*, 21, 972–984.

Dong, C. H. et al. (2006). The negative regulator of plant cold responses, HOS1, is a RING E3 ligase that mediates the ubiquitination and degradation of ICE1. *Proceedings of the National Academy Sciences of the United States of America*, 103, 8281–8286.

Dong, X., Shi, L., Bao, S., Fu, H., You, Y., Ren, Y., & Chen, Z. (2024). Leaf traits of prickly ash and its correlation with ecological and geographical factors of origin. *Scientific Reports*, 14(1), 6276.

Du Plessis, J. (2003). Maize production. Department: Agriculture Republic of South Africa.

Duran Garzon, C., Lequart, M., Rautengarten, C., Bassard, S., Sellier-Richard, H., Baldet, P., Heazlewood, J. L., Gibon, Y., Domon, J. M., & Giauffret, C. et al. (2020). Regulation of carbon metabolism in two maize sister lines contrasted for chilling tolerance. *Journal of Experimental. Botany*, 71, 356–369.

Equiza, M. A., Miravé, J. P., & Tognetti, J. A. (2001). Morphological, anatomical and physiological responses related to differential shoot vs. root growth inhibition at low temperature in spring and winter wheat. *Annals of Botany*, 87(1), 67–76.

Ercoli, L., Mariotti, M., Masoni, A., & Arduini, I. (2004). Growth responses of sorghum plants to chilling temperature and duration of exposure. *European Journal of Agronomy*, 21(1), 93–103.

Farooq, M., Aziz, T., Wahid, A., Lee, D.-J., & Siddique, K. H. (2009). Chilling tolerance in maize: Agronomic and physiological approaches. *Crop and Pasture Science*, 60, 501–516.

Fracheboud, Y., Jompuk, C., Ribaut, J., Stamp, P., & Leipner, J. (2004). Genetic analysis of cold-tolerance of photosynthesis in maize. *Plant Molecular Biology*, 56, 241–253. doi: 10.1007/s11103-004-3353-6

Fracheboud, Y., Ribaut, J. M., Vargas, M., Messmer, R., & Stamp, P. (2002). Identification of quantitative trait loci for cold-tolerance of photosynthesis in maize (Zea mays L.). *Journal of Experimental Botany*, 53, 1967–1977. doi: 10.1093/jxb/erf040

Francis, D., & Barlow, P. W. (1988, January). Temperature and the cell cycle. *Symposia of the Society for Experimental Biology*, 42, 181–201.

Ortiz-Lopez, A., Nie, G. Y., Ort, D. R., & Baker, N. R. (1990). The involvement of the photoinhibition of photosystem ii and impaired membrane energization in the reduced quantum yield of carbon assimilation in chilled maize. *Planta*, 181, 78–84.

Orvar, B. L., Sangwan, V., Ornann, F., & Dhintlsa, R. S. (2000). Early steps in cold sensing by plant cells: The role of actin cytoskeleton and membrane fluidity. *The Plant Journal*, 23, 85–87.

Pal, M., Janda, T., Majlath, I., & Szalai, G. (2020). Involvement of salicylic acid and other phenolic compounds in light-dependent cold acclimation in maize. *International Journal of Molecular Sciences*, 21, 1942.

Peterson, R. L., & Farquhar, M. L. (1996). Root hairs: Specialized tubular cells extending root surfaces. *The Botanical Review*, 62, 1–40.

Pinhero, R. G., Rao, M. V., Paliyath, G., Murr, D. P., & Fletcher, R. A. (1997). Changes in activities of antioxidant enzymes and their relationship to genetic and paclobutrazol-induced chilling tolerance of maize seedlings. *Plant Physiology*, 114, 695–704.

Prasad, T. K. (1997). Role of catalase in inducing chilling tolerance in pre-emergent maize seedlings. *Plant Physiology*, 114, 1369–1376.

Prasad, T. K., Anderson, M. D., Martin, B. A., & Stewart, C. R. (1994). Evidence for chilling-induced oxidative stress in maize seedlings and a regulatory role for hydrogen peroxide. *Plant Cell*, 6, 65–74.

Prasad, T. K., Anderson, M. D., & Stewart, C. R. (1994). Acclimation, hydrogen peroxide, and abscisic acid protect mitochondria against irreversible chilling injury in maize seedlings. *Plant Physiology*, 105, 619–627.

Presterl, T., Ouzunova, M., Schmidt, W., Möller, E. M., Röber, F. K., & Knaak, C. et al. (2007). Quantitative trait loci for early plant vigour of maize grown in chilly environments. *Theoretical and Applied Genetics*, 114, 1059–1070. doi: 10.1007/s00122-006-0499-4

Ramazan, S., Jan, N., & John, R. (2023). Comparative protein analysis of two maize genotypes with contrasting tolerance to low temperature. *BMC Plant Biology*, 23, 183. doi: 10.1186/s12870-023-04198-8

Rape, W. O. (1999). Low temperature affects pattern of leaf growth and structure of cell walls in. *Annals of Botany*, 84, 313–319.

Revilla, P., Malvar, R., Cartea, M., Butrón, A., & Ordás, A. (2000). Inheritance of cold tolerance at emergence and during early season growth in maize. *Crop Science*, 40, 1579–1585. doi: 10.2135/cropsci2000.4061579x

Riley, G. J. P. (1981). Effect of high temperature on the germination of maize (Zea mays L.). *Planta*, 151, 68–74.

Riva, R. L., Escale, B., Giauffret, C., & Périlleux, C. (2016). Maize plants can enter a standby mode to cope with chilling stress. *BMC Plant Biology*, 16, 212.

Roberts, E. H., Hadley, P., & Summerfield, R. J. (1985). Effects of temperature and photoperiod on flowering in chickpeas (Cicer arietinum L.). *Annals of Botany*, 55(6), 881–892.

Rodríguez, V., Butrón, A., Malvar, R. A., Ordás, A., & Revilla, P. (2008). Quantitative trait loci for cold tolerance in the maize IBM population. *International Journal of Plant Sciences*, 169, 551–556. doi: 10.1086/528755

Rodríguez, V., Butrón, A., Rady, M. O., Soengas, P., & Revilla, P. (2014). Identification of quantitative trait loci involved in the response to cold stress in maize (Zea mays L.). *Molecular Breeding*, 33, 363–371. doi: 10.1007/s11032-013-9955-4

Rodríguez, V., Velasco, P., Garrido, J. L., Revilla, P., Ordás, A., & Butrón, A. (2013). Genetic regulation of cold-induced albinism in the maize inbred line A661. *Journal of Experimental Botany*, 64, 3657–3667. doi: 10.1093/jxb/ert189

Rymen, B., Fiorani, F., Kartal, F., Vandepoele, K., Inzé, D., & Beemster, G. T. (2007). Cold nights impair leaf growth and cell cycle progression in maize through transcriptional changes of cell cycle genes. *Plant Physiology*, 143(3), 1429–1438.

Saczynska, V., Kargul, J., & Kaniuga, Z. (1993). Discrimination between chilling-sensitive and chilling-resistant plants based on measurements of free fatty acid accumulation and inactivation of oxygen evolution in aged chloroplasts. *Acta Biochimica Polonica*, 40, 507–513.

Salesse-Smith, C. E., Sharwood, R. E., Busch, F. A., & Stern, D. B. (2020). Increased rubisco content in maize mitigates chilling stress and speeds recovery. *Plant Biotechnology Journal*, 18, 1409–1420.

Sayed, E. G., Desoukey, S. F., Desouky, A. F., Farag, M. F., El-Kholy, R. I., & Azoz, S. N. (2024). Synergistic influence of arbuscular mycorrhizal fungi inoculation with nanoparticle foliar application enhances Chili (*Capsicum annuum* L.) antioxidant enzymes, anatomical characteristics, and productivity under cold-stress conditions. *Plants*, 13, 517.

Sheremet, Y. A., Yemets, A. I., & Blume, Y. B. (2012). Inhibitors of tyrosine kinases and phosphatases as a tool for the investigation of microtubule role in plant cold response. *Tsitologiia i Genetika*, *46*, 3–9.

Shen Y, Wu X, Liu D, Song S, Liu D, Wang H (2017) Cold-dependent alternative splicing of a Jumonji C domain-containing gene *MtJMJC5* in *Medicagotruncatula*. *Biochemical and Biophysical Research Communications*, 474, 271–276

Shi, Y., Ding, Y., & Yang, S. (2018). Molecular regulation of CBF signaling in cold acclimation. *Trends in Plant Science*, 23, 623–637.

Shi, Y., Li, G., Tian, Z., Wang, Z., Wang, X., & Zhu, Y. et al. (2016). Genetic dissection of seed vigour traits in maize (Zea mays L.) under low-temperature conditions. *Journal of Genetics*, 95, 1017–1022. doi: 10.1007/s12041-016-0714-2

Shi, Y., Tian, S., Hou, L., Huang, X., Zhang, X., Guo, H., & Yang, S. (2012). Ethylene signaling negatively regulates freezing tolerance by repressing expression of CBF and type-a ARR genes in *Arabidopsis*. *Plant Cell*, 24, 2578–2595.

Singh, I., Kumar, U., Singh, S. K., Gupta, C., Singh, M., & Kushwaha, S. R. (2012). Physiological and biochemical effect of 24-epibrassinoslide on cold tolerance in maize seedlings. *Physiology and Molecular Biology of Plants*, 18, 229–236.

Soltabayeva, A., Ongaltay, A., Omondi, J. O., & Srivastava, S. (2021). Morphological, physiological and molecular markers for salt-stressed plants. *Plants*, 10(2), 243.

Sowinski, P., Fronk, J., Jończyk, M., Grzybowski, M., Kowalec, P., & Sobkowiak, A. (2020). Maize response to low temperatures at the gene expression level: A critical survey of transcriptomic studies. *Frontiers in Plant Science*, 11, 576941. doi: 10.3389/fpls.2020.576941

Sowinski, P., Rudzinska-Langwald, A., Adamczyk, J., Kubica, I., & Fronk, J. (2005). Recovery of maize seedling growth, development and photosynthetic efficiency after initial growth at low temperature. *Journal of Plant Physiology*, 162, 67–80.

Stockinger, E. J., Gilmour, S. J., & Thomashow, M. F. (1997). Arabidopsis thaliana CBF1 encodes an AP2 domain-containing transcriptional activator that binds to the c-repeat/DRE, a cis-acting DNA regulatory element that stimulates transcription in response to low temperature and water deficit. *Proceedings of the National Academy of Sciences of the United States of America*, 94, 1035–1040.

Strigens, A., Freitag, N. M., Gilbert, X., Grieder, C., Riedelsheimer, C., Schrag, T. A., ... & Melchinger, A. E. (2013). Association mapping for chilling tolerance in elite flint and dent maize inbred lines evaluated in growth chamber and field experiments. *Plant, Cell & Environment*, *36*(10), 1871–1887.

Summerfield, R. J., & Wien, H. C. (1980). Effects of photoperiod and air temperature on growth and yield of economic legumes.

Szalai, G., Majlath, I., Pal, M., Gondor, O. K., Rudnoy, S., Olah, C., Vankova, R., Kalapos, B., & Janda, T. (2018). Janus-faced nature of light in the cold acclimation processes of maize. *Frontiers in Plant Science*, 9, 850.

high-throughput sequencing platforms, coupled with reduced sequencing costs and increased marker density, facilitated genome-wide association studies (GWASs) and the integration of RNA-seq data, enabling the discovery and refinement of candidate genes with higher precision. Conventional breeding has long provided the foundation for LT-tolerant maize improvement, focusing on selecting and recombining naturally resilient genotypes adapted to cooler climates. However, the complexity and polygenic nature of LT tolerance have limited the pace of genetic gains. The advent of molecular breeding approaches, including QTL mapping, marker-assisted selection (MAS), genomic selection (GS), and genetic transformation, has transformed maize improvement strategies, allowing researchers to dissect the genetic architecture of tolerance traits and accelerate cultivar development. More recently, advances in genetic engineering have further expanded the possibilities for maize improvement. Since the introduction of the first genetically modified (GM) maize varieties in the 1990s, biotechnological tools have evolved to target not only biotic but also abiotic stress tolerance, including LT stress.

Taken together, the integration of conventional breeding for adaptive background selection, molecular breeding for genetic precision, and genetic engineering for targeted trait enhancement represents a comprehensive strategy for developing maize cultivars resilient to LT stress. This chapter discusses the current trends, scientific advances, and future prospects in maize LT-tolerant research, highlighting physiological, molecular, and breeding-based innovations aimed at achieving stable and sustainable maize production under fluctuating temperature regimes.

7.2 TRENDS IN MAIZE AND LT TOLERANCE

Over the past decade (2013–2022), considerable progress has been made in understanding maize LT tolerance, as evidenced by the publication of 41 research papers during this period. Among these, 13 studies focused on the antioxidant system, 15 on photosynthetic physiology, and five each on membrane system parameters and hormonal regulation, while eight papers explored osmoregulatory substances. Notably, 12 studies simultaneously investigated more than two physiological indicators associated with LT tolerance, and 11 of these utilized populations rather than a few inbred lines or hybrids for LT-tolerant gene discovery. Consequently, the number of publications on maize cold tolerance during 2013–2022 exceeded the total from the preceding 30 years (Ma et al., 2022). Across all time periods, research most frequently addressed the membrane system, antioxidant system, and photosynthetic parameters, underscoring that LT primarily influences maize through these physiological processes. Moreover, studies focusing on photosynthetic traits and osmoregulatory compounds have shown a consistent upward trend over recent years. Interestingly, the proportion of studies combining multiple indicators (≥2) has remained relatively stable across decades, and cold treatment temperatures have typically ranged from 2°C to 20°C (Ma et al., 2022).

In terms of content, early studies (before 2005) mainly emphasized changes in physiological and biochemical parameters before and after LT exposure, often using a single genotype. Later investigations compared these parameters between LT-tolerant and LT-sensitive maize lines. In tolerant genotypes, higher activity of antioxidant enzymes—such as superoxide dismutase, peroxidase, catalase, and glutathione reductase—was observed, resulting in less damage to the photosynthetic system. Similarly, LT-tolerant materials exhibited higher concentrations of osmoregulatory compounds after exposure to LTs. Although most of these studies evaluated only a few inbred lines or hybrids, the parameters measured provided meaningful insights into maize LT tolerance. Based on such physiological and biochemical responses, several strategies have been developed to enhance LT tolerance in maize. For instance, manipulating unsaturated fatty acid content can decrease electrolyte leakage and preserve cellular membrane integrity under LT stress. Furthermore, exogenous treatments with compounds such as melatonin, chitosan, salicylic acid, and ABA have been shown to stabilize the antioxidant system, reduce photoinhibition, and enhance the accumulation of osmoregulatory substances, including soluble sugars, thereby improving cold tolerance.

While these external treatments offer practical benefits, genetic improvement remains a more sustainable and fundamental approach for enhancing maize cold tolerance. Physiological and biochemical traits measured before and after cold exposure provide critical phenotypic data for identifying the underlying genetic loci controlling chilling tolerance. Based on such phenotypic variation, recombinant inbred lines, $F_{2:3}$ populations, and diploid populations have been developed from tolerant and sensitive parents. Linkage mapping using hundreds to thousands of simple sequence repeat (SSR), restriction fragment length polymorphism (RFLP), or single nucleotide polymorphism (SNP) markers has enabled the identification of QTLs associated with cold tolerance. However, the low marker density and broad confidence intervals (often spanning several megabases) limited these studies to coarse QTL detection. In more recent work (post-2020), RNA-seq data from biparental lines exposed to LTs have been integrated with linkage analyses to pinpoint cold-tolerant QTLs with greater precision. Comprehensive analyses of candidate gene sequences (both nucleotide and amino acid), expression profiles before and after cold treatment, and homologous gene annotations have subsequently led to the identification of multiple candidate genes. These genes exhibit significant sequence variation and expression differences between parental lines and are likely to be key regulators of cold tolerance in maize.

The rapid advancement of high-throughput sequencing technologies, coupled with the complete sequencing and annotation of the maize inbred line B73, has facilitated genome-wide exploration of maize germplasm. Association analysis—originally developed for human disease studies (Thornsberry et al., 2001)—has become a powerful tool in plant genetics, especially for dissecting complex quantitative traits such as LT tolerance. The first GWAS on maize cold tolerance was reported in 2013 (Strigens et al., 2013). Subsequent research applied GWAS to different developmental stages, particularly germination and early seedling growth. The number of SNPs used in these analyses increased dramatically, reaching tens of thousands to millions, thereby enabling high-resolution mapping. Because linkage disequilibrium decays rapidly in the maize genome, the QTL intervals containing significant SNPs are relatively small—often only a few hundred kilobases—allowing more precise localization of candidate genes. After 2016, integrative analytical frameworks emerged, combining association analysis with linkage mapping, RNA-seq data, or photoinhibition, reflected by stabilized Fv/Fm ratios and elevated antioxidant enzyme activity. Overexpression of *ZmMYB–IF35* or *ZmRR1* reduces ion leakage, thereby mitigating cold-induced membrane damage. In *RAF1–LSSS* overexpression lines, photosynthetic parameters (ΦPSII, ΦII, Fv/Fm) show minor reductions, maintaining lower levels of photoinhibition.

Future research directions should focus on the following aspects:

1. Greater emphasis should be placed on functional validation of chilling-tolerance-related genes to unravel key regulatory modules and their upstream and downstream interactions. A complete depiction of the cold signal transduction chain, from perception to response, remains essential. Additionally, exploring cross-talk between cold signaling and other stress pathways such as drought, salinity, and hormone responses will deepen our understanding of the intricate regulatory networks underlying chilling stress adaptation.
2. Large-scale sequencing of diverse maize inbred lines, including ancestral *teosinte* accessions, is necessary to enrich natural variation data. This will facilitate the identification of favorable alleles and illuminate the evolutionary and genetic mechanisms of LT tolerance, including the molecular basis of heterosis in chilling tolerance.
3. Further advancements in maize transformation technologies are crucial, given their genotype dependency. Integrating GS, CRISPR/Cas genome editing, and molecular breeding approaches will enable the pyramiding of beneficial alleles for LT-tolerant genes and their regulatory modules into elite cultivars. Such efforts will expedite the creation of novel LT-tolerant maize germplasm, ensuring high and stable yields under LT stress conditions.

7.4 CONVENTIONAL AND MOLECULAR BREEDING APPROACHES

7.4.1 Conventional Breeding

Traditional breeding methods have long served as the foundation for improving cold tolerance in maize, primarily through the selection and crossbreeding of genotypes that naturally exhibit chilling resilience. Several studies have underscored the significance of conventional breeding in identifying and utilizing cold-tolerant landraces adapted to high-altitude or high-latitude environments. For instance, maize varieties originating from temperate regions have successfully served as donor lines to enhance seedling cold tolerance in tropical maize populations through recurrent selection and hybridization (Zhou et al., 2022). Despite its historical importance and continued relevance, progress in conventional breeding for cold tolerance has been inherently slow due to the complex, quantitative nature of this trait, which is controlled by multiple genes with small additive effects.

Designing a breeding program for improving chilling tolerance requires careful consideration of experimental goals and environmental conditions. Both field and controlled-environment experiments have specific limitations when used independently; hence, a combination of both approaches, conducted across multiple years and locations, is recommended to capture the full extent of genetic variation in chilling tolerance (Frascaroli & Revilla, 2019). The timing and duration of cold stress also play a critical role in determining breeding outcomes. Selection for genotypes capable of acclimating to long-term LTs may yield different results compared to those that can withstand short-term cold snaps under otherwise mild conditions. Importantly, cold stress can adversely affect maize at any developmental stage—reducing germination rates, slowing vegetative growth, or impairing reproductive success. It is noteworthy that improvements in chilling tolerance do not always correlate directly with increased yield, and in some cases, a trade-off between stress tolerance and yield potential may occur (Revilla et al., 2005). Nevertheless, historical yield improvements in maize have been strongly associated with enhanced stress resilience (Tollenaar & Wu, 1999). Therefore, successful breeding strategies must clearly define the growth stage of interest and identify key traits or trait combinations that contribute to cold tolerance. Moreover, resource use efficiency—often masked under optimal conditions—may only become apparent when plants are exposed to chilling stress (Tollenaar & Wu, 1999). In this regard, chilling stress during the reproductive phase remains relatively underexplored, representing an important avenue for future research in the context of increasingly unpredictable climatic conditions. Overall, traditional breeding for abiotic stress tolerance, including cold stress, has achieved only modest progress in maize (Watson, 2019). Several factors have contributed to this limitation. Conventional programs have largely prioritized grain yield as the primary selection criterion, often overlooking specific physiological and adaptive traits that underpin stress resilience. This yield-centric approach has constrained the development of cultivars with robust tolerance to environmental stressors. Additionally, the complexity of genotype-by-environment (G × E) interactions further complicates breeding for cold tolerance, as these interactions obscure the genetic architecture of tolerance traits and reduce selection efficiency. Consequently, developing maize cultivars that maintain stable performance across diverse and fluctuating environments remains a significant challenge.

7.4.2 Molecular Breeding

From the earliest stages of maize improvement, a range of genetic and biotechnological tools have been developed to address challenges related to yield, nutrition, and stress resilience. Foundational discoveries from the Darwinian and Mendelian eras of genetics laid the groundwork for modern molecular breeding by revealing how genetic variation drives crop evolution and inheritance. Molecular breeding focuses on manipulating and selecting genes using techniques such as QTLs mapping, MAS, GS, and genetic transformation. These approaches have transformed crop improvement by deepening our understanding of the genetic basis of complex traits, thereby complementing conventional breeding methods. The integration of molecular markers, gene discovery, and transformation techniques has now become central to plant breeding and crop science research. In particular, combining MAS with traditional breeding allows for the enrichment of desirable alleles, enhancing agronomic, nutritional, and genotypic performance. GS, in turn, enables faster and more cost-effective genetic gains compared to conventional selection (Voss-Fels et al., 2019).

The advent of molecular breeding has markedly enhanced the precision and efficiency of developing maize varieties with improved cold tolerance. Since the introduction of molecular markers and QTL mapping, several studies have identified genomic regions associated with chilling resilience, though their direct impact on breeding outcomes has been limited (Leipner et al., 2008). Reported QTLs for cold tolerance have been linked to physiological traits such as chlorophyll content, photosynthetic efficiency, and early seedling vigor (Jompuk et al., 2005; Presterl et al., 2007; Rodríguez et al., 2014). In a GWAS, Strigens et al. identified 19 QTLs for cold tolerance in maize inbred lines, collectively explaining between 5.7% and 52.5% of phenotypic variance for early growth and chlorophyll fluorescence. Given the highly complex genetic architecture of cold tolerance traits, these authors proposed that **whole-genome prediction** approaches may offer greater potential than classical MAS for improving chilling tolerance in maize.

Collectively, molecular breeding tools—ranging from QTL mapping and MAS to genomic and whole-genome prediction approaches—represent a significant advancement in the quest for cold-tolerant maize. While the translation of genomic discoveries into practical breeding outcomes remains an ongoing challenge, integrating molecular tools with phenomics, bioinformatics, and advanced predictive modeling offers promising opportunities to accelerate the development of resilient maize cultivars suited to cold-prone environments.

7.5 TRANSGENIC/GENETIC ENGINEERING APPROACHES

In addition to traditional and molecular breeding approaches, the last two decades have witnessed a global expansion in the development and cultivation of genetically engineered (GM) maize varieties. Since the approval of the first transgenic maize lines, such as MON 809 and MON810, in the United States in 1996, the number of authorized and cultivated GM maize varieties has steadily increased. According to three major databases, the International Service for the Acquisition of Agri-biotech Applications (ISAAA) lists 240 entries, the Biosafety Clearing House (BCH) lists 298, and the Food and Agriculture Organization (FAO) GM Foods Platform lists 179 entries for GM maize (ISAAA; BCH; FAO). A comprehensive meta-analysis demonstrated that, over two decades of cultivation (1996–2016), GM maize provided multiple benefits for farmers, consumers, and the environment—resulting in higher grain yield, reduced mycotoxin contamination, and fewer impacts on non-target organisms (Pellegrino et al., 2018). This analysis further reported that GM maize represents nearly one-third of all maize cultivated globally. The first generation of GM maize varieties, such as MON809, was primarily designed to combat biotic stressors, including weed competition and insect infestation. These transgenic lines expressed dual traits: herbicide (glyphosate) tolerance (Lee et al., 2002) and lepidopteran insect resistance through the expression of *Cry1Ab* endotoxin. While these varieties addressed pressing biotic challenges, the escalating impacts of climate change now demand genetic solutions targeting abiotic stress tolerance, including cold stress. Molecular and biotechnological tools available today make this goal attainable, though societal mistrust and political barriers continue to pose challenges to the adoption of genetically engineered crops (Malyska et al., 2016).

Genetic modification approaches also provide valuable means to improve photosynthetic efficiency and chilling tolerance in maize. Enhancing Rubisco content and electron transport capacity can substantially increase photosynthetic performance in C_4 plants, as Rubisco plays a pivotal role in recovery following chilling stress (Sales et al., 2021). Overexpression of Rubisco large and small subunits, together with the Rubisco Assembly Factor 1 (RAF1), has been shown to increase Rubisco content by more than 30%, thereby accelerating photosynthetic recovery after chilling exposure (Salesse-Smith et al., 2018). Similarly, the transgenic introduction of chilling-tolerant *PPDK* (pyruvate orthophosphate dikinase) reduced the threshold for cold damage and enhanced photosynthetic rates by 23% under chilling conditions (8°C) (Ohta et al., 2004, 2006). The introduction of the osmoprotectant molecule glycine betaine into maize also improved photosynthesis and reduced chilling-induced injury (Quan et al., 2004).

Although much of the transgenic research has been conducted in other species, these studies offer useful proofs of concept for maize improvement. For instance, the *AlSAP* gene from the grass *Aeluropus littoralis* enhanced photosynthesis and stress tolerance in rice under chilling and other abiotic stress conditions (Ben Saad et al., 2012). Similarly, studies in *Arabidopsis* have shown that *CBF/DREB1* transcription factors play a central role in regulating chilling responses (Miura & Furumoto, 2013). Transgenic expression of *CBF/DREB1* from *Arabidopsis* in tobacco and wheat increased chilling tolerance (Sanghera et al., 2011), while several genes from this family have also been successfully introduced into rice to enhance cold tolerance (da Cruz et al., 2013).

In maize and other crops, activation of latent genes within the genome and understanding their regulatory networks are equally important for improving cold tolerance (Revilla et al., 2005). Transgenic methods can be used to elucidate the function of these endogenous genes. For example, overexpression of a stress-responsive binding factor from the Antarctic grass *Deschampsia antarctica* in rice identified a set of chilling-responsive genes regulated by this factor through RNA-seq analysis (Byun et al., 2018). Furthermore, genome editing technologies such as CRISPR/Cas9 have opened new avenues for introducing beneficial alleles into existing germplasm (Waqas et al., 2021).

Genetic engineering also provides a direct approach to enhance maize cold tolerance by introducing or modifying genes that regulate stress adaptation. One prominent example involves the overexpression of *S-adenosylmethionine decarboxylase (SAMDC)*, a key enzyme in polyamine biosynthesis. Polyamines like spermidine and spermine help stabilize membranes, scavenge ROS, and regulate stress-responsive gene expression (Jiao et al., 2022). Transgenic maize lines overexpressing *SAMDC* exhibit higher polyamine and proline accumulation, elevated antioxidant enzyme activity, and improved seedling-stage cold tolerance. Recent discoveries have also highlighted key regulatory genes involved in cold stress responses. The *COOL1* gene, encoding a bHLH transcription factor, has been identified as a central regulator of maize cold adaptation. Natural variants of *COOL1* from high-latitude maize populations exhibit enhanced interaction with the transcription factor HY5, leading to activation of downstream cold-responsive genes (Song & Cao, 2025). Additionally, phosphorylation and stabilization of *COOL1* by the CPK17 kinase during cold stress

Liu, J., Shi, Y., & Yang, S. (2018). Insights into the regulation of c-repeat binding factors in plant cold signaling. *Journal of Integrative Plant Biology*, 60, 780–795.

Ma, Y., Dai, X., Xu, Y., Luo, W., Zheng, X., Zeng, D., ... & Chong, K. (2015). COLD1 confers chilling tolerance in rice. *Cell*, 160(6), 1209–1221.

Ma, Y., Tan, R., & Zhao, J. (2022). Chilling tolerance in maize: Insights into advances—Toward physio-biochemical responses' and QTL/genes' identification. *Plants*, 11(16), 2082. doi: 10.3390/plants11162082

Malyska, A., Bolla, R., & Twardowski, T. (2016). The role of public opinion in shaping trajectories of agricultural biotechnology. *Trends in Biotechnology*, 34, 530–534. doi: 10.1016/j.tibtech.2016.03.005

Miura, K., & Furumoto, T. (2013). Cold signaling and cold response in plants. *International Journal of Molecular Sciences*, 14, 5312–5337.

Ohta, S., Ishida, Y., & Usami, S. (2004). Expression of cold-tolerant pyruvate, orthophosphate dikinase cDNA, and heterotetramer formation in transgenic maize plants. *Transgenic Research*, 13, 475–485.

Ohta, S., Ishida, Y., & Usami, S. (2006). High-level expression of cold-tolerant pyruvate, orthophosphate dikinase from a genomic clone with site-directed mutations in transgenic maize. *Molecular Breeding*, 18, 29–38.

Pandey, S., Singh, A., Parida, S. K., & Prasad, M. (2022). Combining speed breeding with traditional and genomics-assisted breeding for crop improvement. *Plant Breeding*, 141(3), 301–313.

Pellegrino, E., Bedini, S., Nuti, M., & Ercoli, L. (2018). Impact of genetically engineered maize on agronomic, environmental and toxicological traits: A meta-analysis of 21 years of field data. *Scientific Reports*, 8, 3113. doi: 10.1038/s41598-018-21284-2.

Pesev, N. V. (1970). Genetic factors affecting maize tolerance to low temperatures at emergence and germination. *Theoretical and Applied Genetics*, 40, 351–356.

Presterl, T., Ouzunova, M., Schmidt, W., Moeller, E. M., Roeber, F. K., & Knaak, C. et al. (2007). Quantitative trait loci for early plant vigour of maize grown in chilly environments. *Theoretical and Applied Genetics*, 114, 1059–1070.

Quan, R., Shang, M., Zhang, H., Zhao, Y., & Zhang, J. (2004). Improved chilling tolerance by transformation with betA gene for the enhancement of glycinebetaine synthesis in maize. *Plant Science*, 166, 141–149.

Revilla, P., Butrón, A., Cartea, M., Malvar, R., & Ordás, A. (2005). Breeding for cold tolerance. In: Ashraf M., & Harris P. J. C., eds. *Abiotic stresses. Plant resistance through breeding and molecular approaches*. Boca Raton, FL: CRC Press, 301–398.

Rodríguez, V. M., Butrón, A., Rady, M. O. A., Soengas, P., & Revilla, P. (2014). Identification of QTLs involved in the response to cold stress in maize (*Zea mays* L.). *Molecular Breeding*, 33, 363–371.

Rodríuez, V. M., Butrón, A., Malvar, R. A., Ordás, A., & Revilla, P. (2008). QTLs for cold tolerance in the maize IBM population. *International Journal of Plant Sciences*, 169, 551–556.

Sales, C. R. G., Wang, Y., Evers, J. B., & Kromdijk, J. (2021). Improving C4 photosynthesis to increase productivity under optimal and suboptimal conditions. *Journal of Experimental Botany*, 72, 5942–5960.

Salesse-Smith, C. E., Sharwood, R. E., Busch, F. A., Kromdijk, J., Bardal, V., & Stern, D. B. (2018). Overexpression of Rubisco subunits with RAF1 increases Rubisco content in maize. *Nature Plants*, 4, 802–810.

Sanghera, G., Wani, S., Hussain, W., & Singh, N. (2011). Engineering cold stress tolerance in crop plants. *Current Genomics*, 12, 30–43.

Shi, Y., Ding, Y., & Yang, S. (2018). Molecular regulation of cbf signaling in cold acclimation. *Trends in Plant Science*, 23, 623–637.

Song, X., & Cao, X. (2025). Natural polymorphisms of COOL1 contribute to cold tolerance adaptation in maize. *Science China Life Sciences*, 68(5), 1528–1529.

Strigens, A., Freitag, N. M., Gilbert, X., Grieder, C., Riedelsheimer, C., & Schrag, T. A. et al. (2013). Association mapping for chilling tolerance in elite flint and dent maize inbred lines evaluated in growth chamber and field experiments. *Plant, Cell & Environment*, 36, 1871–1887.

Thornsberry, J. M., Goodman, M. M., Doebley, J., Kresovich, S., Nielsen, D., & Buckler, E. S. (2001). Dwarf8 polymorphisms associate with variation in flowering time. *Nature Genetics*, 28, 286–289.

Tollenaar, M., & Wu, J. (1999). Yield improvement in temperate maize is attributable to greater stress tolerance. *Crop Science*, 39, 1597–1604.

Voss-Fels, K. P., Cooper, M., & Hayes, B. J. (2019). Accelerating crop genetic gains with genomic selection. *Theoretical and Applied Genetics*, 132(3), 669–686.

Waqas, M. A., Wang, X., Zafar, S. A., Noor, M. A., Hussain, H. A., Azher Nawaz, M., & Farooq, M. (2021). Thermal stresses in maize: Effects and management strategies. *Plants*, 10, 2931–2923.

Watson, D. (2019). Adaption to climate change: climate adaptive breeding of maize, wheat and rice. In: Sarkar, A., Sensarma, S., & vanLoon, G., eds. *Sustainable solutions for food security*. Cham: Springer. doi: 10.1007/978-3-319-77878-5_4

Zeng, R., Li, Z., Shi, Y., Fu, D., Yin, P., Cheng, J., & Yang, S. (2021). Natural variation in a type-a response regulator confers maize chilling tolerance. *Nature Communications*, 12(1), 4713.

Zeng, R., Shi, Y., Guo, L., Fu, D., Li, M., Zhang, X., & Yang, S. (2025). A natural variant of COOL1 gene enhances cold tolerance for high-latitude adaptation in maize. *Cell*, 188(5), 1315–1329.

Zhou, X., Muhammad, I., Lan, H., & Xia, C. (2022). Recent advances in the analysis of cold tolerance in maize. *Frontiers in Plant Science*, 13, 866034.

8 Conclusions and Future Prospects

Maize (*Zea mays* L.), one of the most diverse and globally significant cereal crops, has traversed an extraordinary evolutionary and agronomic journey from its tropical origins to successful cultivation in temperate environments. This expansion was achieved through centuries of selection and breeding that have gradually modified its genetic architecture, enabling adaptation to cooler climates. Despite this progress, low-temperature (LT) stress remains a major constraint on maize productivity, particularly in regions where early season chilling or late spring frost coincides with sensitive growth stages, such as germination, emergence, and reproductive development. Understanding how maize responds to LT at multiple biological levels has advanced considerably over the past two decades. Morphological and anatomical modifications, such as enhanced root development and thicker leaf mesophyll, are supported by complex physiological adjustments, including the activation of antioxidant defense mechanisms, osmolyte accumulation, hormonal signaling, and maintenance of photosynthetic efficiency. At the molecular scale, omics technologies—genomics, transcriptomics, proteomics, and metabolomics—have provided unprecedented insight into cold-responsive gene networks, transcription factors, post-translational modifications, and regulatory pathways that underpin stress tolerance. These discoveries have laid a strong foundation for targeted improvement through both conventional and molecular breeding approaches. Nevertheless, substantial challenges remain. The polygenic and environment-dependent nature of cold tolerance complicates selection and breeding, and many key genes identified in model plants are yet to be fully validated in maize under field conditions. Furthermore, the physiological trade-offs between cold tolerance and yield potential necessitate an integrated approach that balances adaptation with productivity. As global climate change continues to alter temperature patterns and seasonal dynamics, the need for maize cultivars capable of maintaining stability across fluctuating environments has become even more urgent.

Future research on maize LT tolerance should focus on combining molecular, physiological, and agronomic approaches to make the crop more adaptable and resilient in different climates. Using systems biology and multi-omics studies can help link genes and proteins to visible plant traits and identify key regulators of LT tolerance. Advanced tools like CRISPR/Cas genome editing, genomic selection, and marker-assisted breeding can speed up the development of maize varieties that stay productive even under LTs. Exploring the genetic diversity in wild relatives and traditional landraces can provide new genes that improve stress resistance. Field-based physiological and modeling studies are also needed to connect laboratory findings with real-world performance. In addition, climate-smart practices such as seed priming, balanced nutrition, and the use of helpful soil microbes can reduce the effects

DOI: 10.1201/9781003634027-8

of early season cold and support healthy crop establishment in cooler and changing environments. Collectively, these strategies hold the potential to strengthen maize resilience to LT and other co-occurring abiotic stresses. Continued collaboration across disciplines; plant physiology, molecular genetics, bioinformatics, and climate science; will be critical to translate laboratory findings into field-ready solutions. The integration of classical breeding wisdom with cutting-edge biotechnology promises not only to expand maize's ecological boundaries but also to safeguard global food security in an era of climatic uncertainty.

D

E

H

I

K

L

T

U

V

W